The Test and Launch Control Technology for Launch Vehicles

Zhengyu Song

The Test and Launch Control Technology for Launch Vehicles

国防工业出版社
National Defense Industry Press

Springer

Zhengyu Song
China Academy of Launch Vehicle
Technology
Beijing
China

ISBN 978-981-10-8711-0 ISBN 978-981-10-8712-7 (eBook)
https://doi.org/10.1007/978-981-10-8712-7

Jointly published with National Defense Industry Press, Beijing

The print edition is not for sale in China Mainland. Customers from China Mainland please order the print book from: National Defense Industry Press, Beijing

Library of Congress Control Number: 2018934445

Printed on acid-free paper

This Springer imprint is published by the registered company Springer Nature Singapore Pte Ltd.
part of Springer Nature
The registered company address is: 152 Beach Road, #21-01/04 Gateway East, Singapore 189721, Singapore

Preface

The main function of a rocket is to fly reliably and stably and deliver the payload into orbit accurately. Although this should be achieved by the excellent performance of onboard systems (OBSs), ground test and launch control systems (TLCSs) also play vital roles to ensure success. On the one hand, the TLCS provides comprehensive checks and verification for the rocket's function and performance, minimizing technological risks in advance; on the other hand, the TLCS itself is an important part of the prelaunch facilities, critical to completing the task reliably, and particularly for launch site safety. To cope with the increasing competition in the launch service market, many countries have recognized TLCS contributions. With the intensive launch activities nowadays, it is widely accepted that improving efficiency, reducing costs, and getting into and out of space responsively, while still ensuring high reliability and safety, have become the major competitiveness factors in the aerospace industry.

This book is written in the context explained above. It is divided into seven chapters, focusing on the design of the launch vehicle ground test and launch control system, giving consideration to the equipment level tests and the simulation tests, and conducting a special discussion on the responsive test and launch control technology. The book comprehensively reflects China's achievement and the latest progress in these fields.

The first chapter introduces TLCS' development worldwide and the status of related technologies in China. The corresponding technologies of the NASA-led project, ESA Ariane 5 launcher, and JAXA Epsilon rocket are introduced as key points of the present situation of the US, Europe, and Japan. Among them, the responsive test and launch control is alleged as the main innovation of Epsilon, the Japanese newly developed small solid rocket. At the end of this chapter, the Chinese TLCS development is briefly reviewed and prospected.

The second chapter introduces the launch vehicle's electrical system testing technology in general. First of all, it sorts out the current testing technologies from various viewpoints and then proposes different corresponding classification methods. The chapter then presents a brief introduction from four aspects of test activities: Equipment level, system level, launch site, and software testing.

Subsequently, test system design is also introduced in terms of the test requirements, including basic test system development processes, design for test (DFT) technology, typical testing equipment with various bus technologies, etc. At the end of this chapter, the future testing technology development is discussed.

The third chapter introduces the equipment level test technology, which incorporates functional/ performance and reliability testing. Considering the features and complexity of the control system equipment functional test, it introduces two equipment categories: Controllers, including the OBC, and inertial devices, including the IMU and other devices composed of accelerometers and rate gyros. It also considers reliability testing, including reliability growth and enhancement tests. At the end of this chapter, some cautions on the product verification (acceptance) tests are analyzed and explained.

The fourth chapter introduces the system level test technology, which incorporates the subsystem and the system level test. The former test is a functional static test where the notion of "subsystem" is relative. In general, the control system itself is a subsystem but the "subsystem" here refers to a small system composed of multiple devices. The subsystem test must cover the functional inspection of all the onboard products under the system connection state, and the test items should be adapted to the implementations on the launch site. The system level test is the flight simulation with a complete LV electrical system, mainly referred as the general check in China and is performed hierarchically. Through the test, the LV avionics performances, the coordination between the OBS and the TLCS, along with the electromagnetic compatibility between the rocket and the ground system, are being assessed. For the newly developed rocket, the engine and propellant system tests are very necessary and are also introduced as system level test.

The fifth chapter discusses the simulation test technology, which is another kind of system level test. It first reviews the simulation technology in the aerospace industry, with particular attention to LVs, and then discusses the simulation testing for LV control systems along with the basic simulation testing principles. Subsequently, it briefly introduces the modeling, simulation process, and the boundary conditions commonly applied for the LV simulation analysis. The modeling relates to the model design of the control system avionics and the LV motion, including the small deviation motion and 6DOF simulation model. The boundary simulation conditions reflect the evaluation criteria on the system performance.

The sixth chapter introduces the launch control technology, taking the widely used remote launch control technology as the basis. The launch control process, the front-end and back-end facilities, and the network systems are being discussed. Because of the importance of the launch control technology, the safety levels of launch control related software are higher than other ground test software. So under this background, the chapter introduces the ground TLCS software system, including software reliability design, and the corresponding technologies unique to the ground TLCS software.

The seventh chapter introduces the responsive test and launch control technology, focusing on the means to improve testing efficiency, simplify operations, and adopt intelligent data analysis to reduce on-site technical support personnel. It reflects the latest research results in this field, and the experience gained in the practical use of other countries' launch vehicles is also supplemented in this chapter.

Due to the space limitation, this book cannot include all aspects of launch vehicle TLCS. It starts with the discussion from the traditional launch and control system, taking into account the electrical equipment level test, engine and propellant system test, as well as the simulation test. The book does not give detailed and specific hardware or software design, but it expatiates mainly from the aspects of testing principle, purpose, process, etc., and lets the readers grasp the nature of test and launch control technologies.

The compilation of this book has been helped by many colleagues, especially Mr. Li Xinming. The intended audiences of this book are engineers of launch vehicle TLCS and avionics system, and operation staffs on the launch sites.

Beijing, China Zhengyu Song
January 2018

Contents

Brief Introduction

This book sums up systematically the author's research work in launch vehicle test and launch control system (TLCS), and the book is divided into seven chapters, where the overall electrical system test, equipment level test, system level test, simulation test, and launch control technology are introduced. In the end of the book, the author discussed special subject for responsive test and launch control, which are essential to the increasing demands for responsive, reliable, and economical to and from the space. There are many best practices in the book, which are benefit for better understanding of these technologies.

This book can serve as references to designers of TlCS, overall and avionics system, and operators in launch site. It also has certain reference value to the test engineers, and other spacecraft designers.

Chapter 1
Introduction

The main function of a rocket is to fly reliably and stably, and deliver the payload into orbit accurately. Although this should be achieved by excellent performance of onboard systems (OBSs), ground test and launch control systems (TLCSs) also play vital roles to ensure success. The TLCS provides comprehensive checks and verification for the rocket's function and performance, minimizing technology risks in advance. This is an important part of the prelaunch facilities, critical to completing the task reliably, and particularly for launch site safety. To cope with increasing competition in the launch service market, many countries have recognized TLCS contributions. NASA estimated that ground operations (including, but more than, TLCS) account for 40% of total project cost, so simple and cost-effective operations are essential [1]. With intensive launch activities in China, it is widely accepted that improving efficiency, reducing costs, and getting into and out of space responsively, while still ensuring high reliability and safety, have become major competitiveness factors in the aerospace industry.

This section introduces TLCS developments worldwide. System level testing and launch control are discussed in subsequent sections.

1.1 United States Check and Launch Control Systems

1.1.1 Launch Control System

Many US launch facilities are under the control of NASA, including Kennedy Space Center in Florida, Johnson Space Center in Texas, and Marshall Space Flight Center in Alabama. Launch operations for United Launch Alliance (ULA) rockets are conducted from Cape Canaveral Air Force Station Florida, and Vandenberg Air Force Base California.

Z. Song, *The Test and Launch Control Technology for Launch Vehicles*,
https://doi.org/10.1007/978-981-10-8712-7_1

Taking the space launch system (SLS) of NASA as an example, launch day operations are scheduled as follows [2].

- Launch control center (LCC) is in Kennedy Space Center (KSC), and the tasks of the launch operations team execute countdown timeline, monitor launch commit criteria, and decide go/nogo for launch.
- Mission control center (MCC) is in Johnson Space Center, and the flight control team is responsible for ascent monitoring, orbit operations, crew communication, and vehicle command.
- SLS engineering support center (SESC) is in Marshall Space Flight Center, and there are two teams.
  - SLS engineering support team monitors prelaunch activities, and monitors launch commit criteria.
  - SLS flight operations support team is responsible for ascent/on-orbit monitoring and supports anomaly resolution.

SESC also provides technical support for LCC and MCC through voice, data, and video and receives technical support from remote (contractor) sites.

Kennedy Space Center is in charge of rocket launch [3] and is the focus of this section. The infrastructure, i.e., the launch processing system (LPS), was designed primarily for test and launch control for the space shuttle. LPS comprises computers, data links, displays, controllers, hardware interface units, and software, integrated through the KSC network. LPS was first activated in 1979 (STS-1, Orbiter Processing Facility) and completed its first manned flight in 1981. It was upgraded in 1996, and initially called the national launch processing system, which was later changed to the check and launch control system (CLCS).

The main reason was to replace outdated LPS technology, using commercial off-the-shelf (COTS) products, and CLCS adopted a distributed and scalable architecture with robust fail-safe measures and improved failure tolerance. It provided significantly more data and information for the terminals with reduced operational and maintenance costs.

Under these design concepts, CLCS consists of four main components.

1. Simulation system. Including various mathematical models, such as shuttles, solid boosters, ground support equipment (GSE), external tanks, payloads, etc., the simulation system provides diagnostic and validation capabilities for real-time processing systems.
2. Real-time processing systems (RTPS). These are used for instruction interaction with terminal products (i.e., onboard equipment, etc.) and information monitoring.
3. Data center. This is the knowledge base (KB) for RTPS software and data records of the whole system.
4. Business and support information service. This is designed for network communication with other subsystems.

NASA proposed a new architecture for future launch control when developing the Ares rocket, based on the CLCS but incorporating new aspects of information technology [4]. The total launch control system is composed of three major parts:

- launch site command and control system, including operations support facilities, control room, vehicle and GSE gateway, which are connected through a message bus;
- advisory/monitor only support facilities, including advisory and monitor workstations; and
- mobile launch pad, where industrial controllers and programmable logic controllers (PLCs) are integrated and installed.

1.1.2 System Testing

The US military has clear requirements for rocket, spacecraft, and upper stage testing [5–7], and are classified as unit, subsystem, or vehicle tests. Unit tests are similar to equipment tests discussed in Chap. 3. Vehicle tests can be performed at the manufacturer's factory or launch site, and include prelaunch validation and follow-on operational tests.

Follow-on operational tests are related to payload operation and are not relevant to the rocket systems.

Prelaunch tests verify the rocket condition and are similar to normal Chinese practice where the rocket is checked in the assembly facility or validated after transport to the launch site. They are divided into two stages [6]: system integration tests (called combined or integration tests in China), which check the rocket itself; and initial operational tests (called regional joint walkthrough in China), which check compatibilities between the rocket, ground systems (GSs), TT & C communication system, and payload.

The main focus of prelaunch tests include

- functionality, including electrical system, structural system and valve leakage, mechanism operability, etc.;
- propulsion system leaks and functionality;
- critical GSE related to launch control; and
- onboard equipment compatibilities, including electromagnetic compatibility, wireless command and communication link reliability, etc.

NASA has specific demands for its own projects, additional to the US military requirements. For example, in the Constellation program [8], rocket testing adopted the V model. User requirements were divided into architecture, system, element, subsystem, and device categories from the top to bottom layer. Then requirement analysis is independently verified for each layer whether the right questions and suitable solutions were proposed. Verification and integration plans were formulated for each layer to verify that corresponding user requirements were met and were in accordance with upper level requirements.

Engine tests are essential for any newly developed rocket, e.g., the Ares rocket main thrust system [9], and the avionics, thrust vector control, pressurizing, and propellant feed systems can also be investigated during engine testing. The integrated vehicle ground vibration test was devised for the Ares rocket, and used for modeling vehicle dynamics.

Many tests are performed during development and need not be repeated once they have provided the desired results. However, integration tests for the avionics system are important prelaunch checks for each launch and considered to be final verification and validation of the rocket.

For the SLS project, NASA built an integrated avionics system testing facility to verify, validate, and confirm the avionics system sequentially [10, 11]. The tests were performed in three phases, and the facilities used in these phases were configured accordingly.

1. SLS software development facility (SDF). The algorithm development environment is provided, including GNC algorithm and vehicle dynamics.
2. SLS element development and integration laboratory, comprising upper stage system integrated test facilities and hardware-in-the-loop (HWIL) test facilities for the first stage and engines. NASA has always stressed the importance of HWIL simulation [12].
3. SLS system integration laboratory (SIL). Avionics of the first and upper stages, engine control devices, launch control system, and crew devices (or corresponding simulators) are integrated for testing.

Tests that incorporate the rocket, payload, and ground, mission, and TT&C network systems are called flight elements integration tests (FEITs) [8], which are the same as the US military prelaunch validation tests discussed above. If a project involves coordinating multiple spacecrafts and rockets, whole system testing is known as multielement integrated test. Compared to FEIT, different space vehicles, such as manned and cargo launch vehicles (LVs), Earth departure stage, and lunar surface access module, are involved.

1.1.3 Fault Diagnosis and Health Management

Automatic fault detection, isolation, and reconstruction (FDIR) has been studied to improve testing efficiency, and applied for launch control systems in the US, implemented as expert or health inspection systems. Expert systems are an important tool for studying artificial intelligence, which began in the 1960s and developed rapidly during 1970–1980s, and have been widely used in industrial and national defense fields, etc. In the 1980s, expert systems were applied to control systems, mainly for fault diagnosis, etc. The technology was also studied in the context of ground test systems for the space shuttle [13]. These systems originally depended on analysis and monitoring by engineers but now employ expert systems.

The Constellation program integrated the FDIR prototype developed in NASA's exploration technology development program into the Ares rocket GS for fault diagnosis, making wide use of functional failure models. A fluid/hydraulic model was also incorporated to improve propellant system diagnosis efficiency.

The US aerospace industry has gradually integrated fault diagnosis into the health management system (HMS), which affects CLCS in two ways [14, 15].

1. FDIR is independent of CLCS. The original system software architecture is retained, and configured with a model-based inference engine (TEAMS) and data-driven (self-learning) inference engine (IMS). FDIR can obtain data through the data bus and application service programs and perform analyses.
2. Embedded or integrated FDIR (IFDIR). FDIR is integrated in three functional areas: FDIR server partition, with TEAMS and IMS installed; application server partition; and display server partition.

Although a great deal of FDIR and integrated vehicle health management (IVHM) research has been performed in the US, only three tools, i.e., SHINE, TEAMS, and IMS have been effectively introduced [14–20]; and SHINE mainly provides a reference for TEAMS fault simulation.

1.2 European Launch Control System

This section discusses the Ariane and VEGA rocket systems.

1.2.1 Ariane 5 Rocket

The Ariane rockets have three launch sites, ELA-1, ELA 2, and ELA-3 [21]. ELA-3 is used for launching Ariane 5, and compared to ELA-1 (Ariane 1,2,3) and ELA-2 (Ariane 4), operations in the launch area have been dramatically reduced, going directly to preparation phase eight hours before launch. ELA-3 includes two functional areas as follows.

1. Preparation zone

   The preparation zone includes the launcher integration (BIL), final assembly (BAF), and launch control center (CDL) buildings.

   The rocket stages are integrated onto the launch pad in the BIL, and payloads are mounted in the rocket in BAF, along with tests to simulate launch to inspect safety controls and telemetry stations. The CDL includes two halls for test and launch control, three control rooms for payloads, and offices. The computers in the control rooms are connected to devices in BAF, BIL, and the launch pad to complete the TLCS.

2. Launch zone
   The launch zone provides launch services for the rocket and launch pad, including fuel filling for core stages, water injection, noise reduction, etc. Electrical and fluid interfaces to the launch pad and corresponding facilities are also provided in a nearby building.

The TLCS and whole GS failure rate is strictly controlled based on previous testing of other practices, particularly during preparation and examination phases before launch to improve availability, safety, reliability, and maintainability; and optimize operation costs. The TLCS comprises two logical sections: front end and back end. Front-end facilities are called processors, and are distributed among the BAF, BIL and launch pad, while the back-end facilities are in the control center. The various facilities can be classified into four categories:

1. check systems for general-purpose equipment, employed to monitor electrical sources, air condition, burning, gas or fluid leakage, etc.
2. operational control center, including processors, execution units, and network interfaces;
3. check systems for the upper stage; and
4. check systems for the payload.

Failure-tolerant or redundant technologies are widely adopted in the Ariane 5 rocket TLCS, mainly following the master–slave topology [22].

1.2.2 Vega

Front-end and back-end facilities have been applied for many launchers. For the VEGA launcher, these facilities were called low-level (LLCS) and high-level (HLCS) control systems, respectively [23]. The HLCS is mainly devoted to operator interaction, allowing automatic procedures and transmitting commands to the LLCS. The LLCS is composed of modular subsystems interfaced with the launcher or ground segments, allowing parameter acquisition, command transmission, and launcher equipment powering.

The architecture contains a test configuration system, mainly dedicated to configuration and validation of the test environment and parameters, and a postprocessing system to analyze and postprocess the data collected during test sessions and produce reports. Both HLCS and LLCS are defined as the test execution system. The LLCS is configured with checking equipment (SCOE in French) and avionic simulators for specific test scenarios. The SCOEs, which include the power supply (PS SCOE), 1553B data bus (1553/SPY SCOE), fluid control (Fluid SCOE), telemetry processing (TM SCOE), and electrical wiring (Wiring SCOE), are similar to the Ariane 5 processors. The avionic simulators, including pyrotechnics, telemetry sensor, upper stage composite, and multifunction unit (similar to onboard computer [OBC]) simulators, are used for load and signal stimulations.

1.3 Japanese Automatic Checking and Rapid Launch Technology

Traditional LVs required many devices for ground inspection prior to launch, and hence significant time and manpower. Many components also had to be manually assembled at the launch center. Consequently, the Japanese Epsilon rocket made vehicle performance checks autonomously onboard to greatly reduce the volume of work and manpower required for ground facility and launch operations. Ultimately, engineers will be able to check and control rockets anywhere in the world using a laptop computer through the Internet, and the launch control system will not be required to be at the launch site anymore [24].

The Epsilon avionics system is divided into five aspects: thrust vector control; navigation and control; power supply; telemetry and communication; and onboard test equipment, such as responsive operation support equipment (ROSE) and miniature check equipment for ignition circuits (MOC) [24, 25]. ROSE is an artificial intelligence system, incorporating ROSE-S and ROSE-M, where ROSE-S is installed on every stage for power control, status monitoring and acquisition; and ROSE-M is installed on the second stage, in addition to ROSE-S, for status monitoring, emergency abort control, data hub, and communication between OBS and GS. MOC is also installed at all stages for ignition circuits checking, and takeoff and stage separation simulation, as required during testing or operation. Data collected by ROSE-S and MOC are sent to ROSE-M and then relayed to GS. MOC is designed to be removed from the rocket before liftoff and can be employed for subsequent launches. These design aspects significantly reduce manpower and operation time.

The Epsilon rocket has also adopted automated testing and diagnosis solutions [26]. The test plan, steps, instructions, and responses are stored in a database (DB). Instructions are read from there and executed, and test data are collected through ROSE. Failure modes are analyzed to locate faults, and the corresponding monitor (or test) items and waveforms under these modes comprise an available KB in the DB. When something unusual occurs, the test item waveform(s) are compared with those stored in the KB, and the fault location and type suggested.

1.4 Chinese Test and Launch Control Technology

Test and launch control technology development in China has benefited from long-term independent research [27]. Manual operations, such as manually reading meters and controlling switches, were very common in the early days of the Chinese aerospace industry. Automatic testing commenced in the 1960s, with programmable circuits, test circuits, excitation signal sources, and sampling switches being gradually adopted. An onboard test scheme was adopted after the first successful long-range flight of the Long March (LM) launcher in 1971, with programmable controlled circuits realized by OBC. Subsequently, the systems have evolved to be capable of checking the equivalent coefficients of inertial devices, transfer coefficients, and dynamic stability of control laws.

The LM launcher TLCS has been inevitably affected by innovation in electrical and electronics technologies. Four institutes in the Chinese aerospace industry collaborated to develop computer automatic measurement and control modules, which have been widely adopted for the LM launchers (such as CZ-2C, CZ-2E, etc.).

Launching is a high-risk activity. Engineers have carefully investigated accidents caused by poor product quality as the Chinese space industry has grown, and realized that the control center should be far from the launch area to ensure safety. Consequently, the Chinese manned spaceflight project TLCS adopted the front-end and back-end configuration, with front-end facilities installed underground in the launch area, near the launch pad, and back-end facilities placed in the control center, which was 2 or 3 km from the launch pad and out of the risk zone if the launcher exploded. Front-end equipment is automatically controlled based on commands received from the control center, or manually by underground operators if necessary. Forty minutes before launch, most operators in the launch area are evacuated.

Various test modules based on standard buses have been widely used in the industry, including VXI, PXI, etc. Thus, OBC-controlled test schemes have been gradually replaced by modular on-ground test facilities, particularly for flight simulation of the entire avionics or electrical system. Tests are mainly static and functional.

The TLCS is a vital part of launch services but initially TLCS focused only on technologies required to check OBSs, and ignored the basic requirements of launch service. These early systems were often referred as an "expert system", in the sense that only experts could operate them. This situation has significantly changed after years of efforts to improve the level of automation and operation convenience, and test and launch control activities are no longer constraints for on-site operations. However, demand is always increasing and promotes continuous development.

References

1. NASA Technology Roadmaps TA 13 (2015) Ground and launch systems, NASA, May 2015 Draft
2. Watson MD, Epps A, Woodruff V (2014) Launch vehicle control center architectures. In: SpaceOps 2014 conference, Pasadema, USA, 5–9 May 2014
3. Hurt GR (2007) History of command and control at KSC. NASA Kennedy Engineering Academy Series, KSC-2007-198
4. Ferrell B, Lewis M, Perotti J et al (2010) Usage of fault detection isolation & recovery (FDIR) in constellation (CxP) launch operations. In: SpaceOps 2010 conference, Hunstville, Alabama, USA, 25–30 April 2010
5. MIL-STD-1540C (1994) Test requirements for launch, upper stage, and space vehicles

6. MIL-STD-1833 (1989) Test requirements for ground equipment and associated computer software supporting space vehicles
7. Perl E (2006) Test requirement for launch, upper-stage, and space vehicles. The aerospace corporation, TR-2004(8583)-1 REV. A, 6 Sep 2006
8. Strong E Test & verification approach for the NASA constellation program[DB/OL]. Constellation program office, Johnson Space Center. http://ntrs.nasa.gov/archive/nasa/casi.ntrs.nasa.gov/20080009753.pdf
9. Cockrell C (2008) Integrated system test approaches for the NASA Ares I crew launch vehicle. In: 59th IAF congress, Glasgow, United Kingdom, 3 Oct 2008
10. NASA (2013) Space launch system program (SLSP) integrated avionics test facilities (IATF) verification, validation, and accreditation (VV&A) plan. NASA, SLS-PLAN-137
11. George C SLS system integration lab and thrust vector control test lab at marshall center's propulsion laboratory. NASA, NASA facts, FS-2014-03-17-MSFC, G-53230
12. Tobbe P, Matras A, Walker D et al (2009) Real-time hardware-in-the-loop simulation of launch vehicle. In: AIAA modeling and simulation technology conference, Chicago, UAS, 10–13 Aug 2009
13. Semmel GS, Davis SR, Leucht KW et al (2006) Space shuttle ground processing with monitoring agents. IEEE Intell Syst 21(1):68–73
14. Schwabacher M, Martin R, Waterman R et al (2010) Ares I-X ground diagnostic prototype. In: Infotech@Aerospace conference, Atlanta, USA, 20–22 Apr 2010
15. Spirkovska L, Iverson DL, Hall DR et al (2010) Anomaly detection for next-generation space launch ground operations. In: AIAA SpaceOps 2010 conference, Huntsville, USA, 25–30 Apr 2010
16. Kurtoglu T, Johnson SB, Barszcz E et al (2008) Integrating system health management into the early design of aerospace systems using functional fault analysis. In: International conference on prognostics and health management, Denver, USA, 6–9 Oct 2008
17. Ferrell B, Oostdyk R (2010) Modeling and performance considerations for automated fault isolation in complex systems. In: IEEE aerospace conference proceedings, Big Sky, USA, 6–13 Mar 2010
18. Johnson SB, Gormley TJ, Kessler SS et al (2011) System health management: with aerospace applications. Wiley, USA
19. Iverson DL (2004) Inductive system health monitoring. In: Proceedings of the international conference on artificial intelligence, IC-AI '04, Las Vegas, USA, 21–24 June 2004
20. Biswas G, Mack D, Mylaraswamy D et al Data mining for anomaly detection. NASA, NASA/CR-2013-217973
21. de Dalmau J (1994) The Ariane launch facilities. ESA bulletin, 1994.8
22. Dega JL (1996) The redundancy mechanisms of the Ariane 5 operational control center. In: Proceedings of annual symposium on fault tolerant computing, Sendai, Japan, 25–27 Jun 1996
23. Angioli E, Chicarella C, Ciaccini M et al (2012) The EGSE and CCV for ESA VEGA launcher. In: SESP 2012: simulation and EGSE facilities for space programmes, Noordwijk, Holland, 25–27 Sep 2012
24. Masutaka H, Nankai N, Inoue T et al (2012) The avionics system for the epsilon launch vehicle. JAXA, STCP-2012-011 (in Japanese)
25. Morita Y (2012) A year to launch: Japan's epsilon launcher and its evolution. In: 63rd international astronautical congress, 2012, IAC-12-D2.1.8, Naples, Italy, 1–5 Oct 2012
26. Epsilon火箭发射倒计时[EB/OL]. JAXA. http://www.isas.jaxa.jp/j/column/ep_countdown
27. Song ZY, Zhang L (2007) Development and prospect of Chinese launch vehicle testing technology. In: 1st high-level forum on national defense science and technology industry experiment and testing technology, Beijing, China, 21–22 Sep 2007 (in Chinese)

Chapter 2
Test Technology for Launcher Electrical Systems

Testing and checking continue for the whole launcher life cycle, and it is difficult to introduce all issues related to testing. Test content is very complicated and involves many aspects. Many test activities are performed during the development process and are not repeated in follow-up services such as structural strength checks, whereas avionics or electrical system tests are implemented before every launch. Therefore, this section introduces electrical system test techniques.

Even for the electrical system, comprehensive tests differ for different rockets and application requirements. Hence, test activities are also scheduled differently, e.g., reliability requirements have a great influence on required testing. This section first discusses test activities or techniques from different viewpoints, to provide a better understanding of the field; then, presents a brief introduction from the four aspects of test activities: equipment level, system level, launch site, and software testing. Detailed content will be discussed in the following chapters.

Test system design is also introduced in terms of the test requirements, including basic test system development processes, design for test (DFT) technology, typical testing equipment, etc. In our review of test equipment, we focus on test modules based on standard buses and environmentally adaptable design.

Finally, we discuss future test technology development.

2.1 Test Technology Hierarchy

Rocket electrical systems are mainly composed of two subsystems: control (GNC) and telemetry systems. The latter collects onboard signals (i.e., telemetry data), transmits them to the GS, and receives control commands from the control center. The former is the key nerve center or control center for the flight and the focus of prelaunch checks [1]. The engine, GSE, etc., all have their own testing requirements. A complete rocket test incorporates step-by-step integration testing from subsystem to overall system, and scientific and rational for test and evaluation

Z. Song, *The Test and Launch Control Technology for Launch Vehicles*,
https://doi.org/10.1007/978-981-10-8712-7_2

methods can significantly shorten the development cycle, improve test efficiency, and reduce application risk.

After 50 years of growth, the Chinese aerospace industry has developed its own test technical system, which has a vital role in ensuring successful launches. Figure 2.1 shows that rocket test items can be classified depending on where and what test activities are required [2–4].

The equivalent state test is used for self-testing of ground devices. System testing is mainly performed in the integration laboratory for each subsystem, and integration (matching) tests between systems are usually performed in the assembly department. The general check refers to the joint check of the whole electrical system to simulate the flight process. In most situations, the separation connectors between different stages remain unchanged, but sometimes these connectors are separated to simulate stage separation processes. Batteries might also be used during the general check to check power switching from ground power supplies to onboard batteries. Most launch site tests are conducted in the technical center.

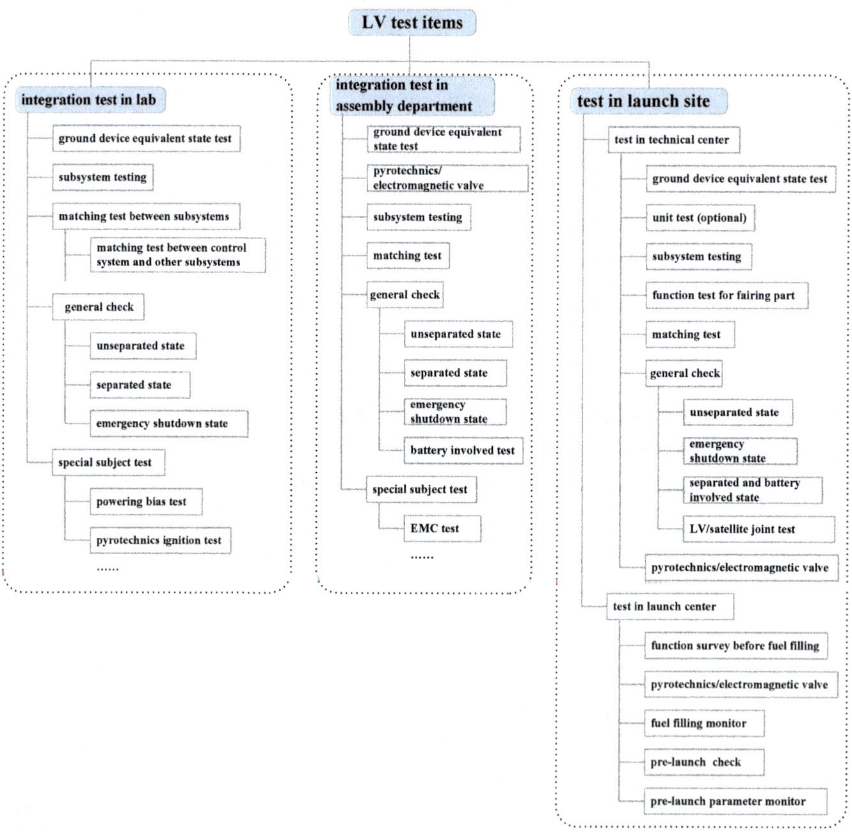

Fig. 2.1 Launch vehicle test items

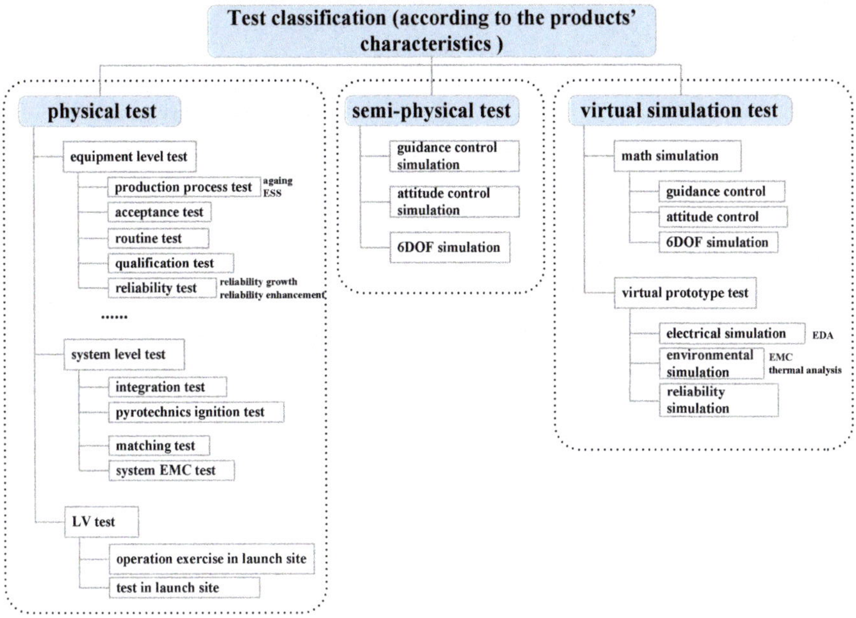

Fig. 2.2 Test items classified according to the products' characteristics

Considering the products' characteristics, test items can be classified as physical, semi-physical, and virtual simulation tests, as shown in Fig. 2.2.

As the name suggests, physical tests are made on real products. Usually, a single prototype is selected for qualification inspection, to verify that the product design not only meets the specifications but also has sufficient design margin. Qualification test conditions are very strict, determined by predictive environmental estimates based on 99% probability and 90% confidence (P99/90) [5], or exceeding the acceptance test conditions with some allowance. If the prototype passes the qualification test, the design is recognized as having been fully examined, with guaranteed design reliability to withstand the worst expected environmental conditions and a certain number of acceptance tests. Products used for qualification testing cannot be subsequently used for flight operations, due to concerns their service life may have been exhausted.

Acceptance tests are performed on every product to identify potential defects during production, including defects related to manufacturing, assembly, components and materials, and to screen early faults. In general, test conditions are specified from predicted environment estimates based on 95% probability and 50% confidence (P95/50), i.e., the highest expected environmental conditions. In addition, acceptance test conditions should also be strict enough to simulate defects.

Test items can be classified based on the products' features being evaluated, as functional (performance), reliability, and environmental adaptability testing, as shown in Fig. 2.3.

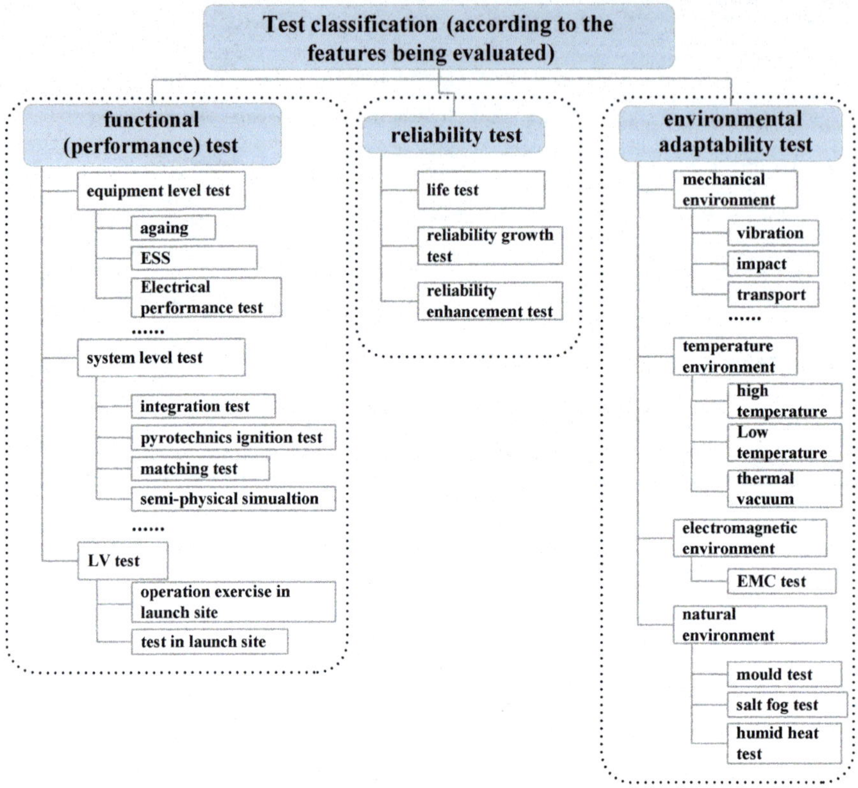

Fig. 2.3 Test items classified according to the features being evaluated

Environmental adaptability testing is required for onboard products, and this is the main difference from the test requirement for ground equipment. Using the mechanical environment as an example, onboard products will experience harsh mechanical environments during liftoff, shutdown, and separation. Although their duration is short, they have significant influence on product structural reliability, as well as normal operation throughout the whole flight process.

2.2 Introduction of Test Items

To facilitate further discussion, this section provides a general introduction to test content from four aspects: equipment level, system level, launch site, and software testing.

2.2.1 Equipment Level Test

Many standards cover equipment level testing [6, 7], and each rocket requires its own special requirements. Equipment level tests include four categories as follows, with Table 2.1 showing test items included in categories 1–3.

1. Production tests, such as screening, acceptance, etc. tests. Every product is tested.
2. Selected routine tests. One or more items are selected from a batch of products to test.
3. Qualification tests. Usually only implemented once if the product technical status is unchanged.
4. Reliability tests. Used to identify product weaknesses, improve design, and recheck to enhance reliability. Tests are usually scheduled one or two times during the development phase of the rocket.

These activities are planned at various development phases depending on possible product impacts. Selected routine testing is a sample inspection to assess batch production quality consistency.

Qualification tests assess design and product quality, and although the acceptance and routine tests can play a certain effect for qualification, but they are not enough, other test items are required, and this is what the qualification tests aim for. The qualification tests are usually performed only once because once passed, the test result represents all products with the same technical state or only few changes of the state.

The implementation order of the test items within each category is arranged depending on the effects on the unit under test (UUT), to ensure prior test items on the UUT do not affect subsequent test items, otherwise new UUTs are required for these test items. Electrical performance testing is important to check avionics product functions and performance, and is usually implemented at room temperature. However, the electrical performance testing is also performed throughout all other tests, where the UUT should be powered on, and the test results are compared with those from room temperature tests.

Here, we consider some mechanical environment test items as the examples [8] for more detailed introduction.

1. Acceleration test

   Acceleration tests simulate stable or quasi-stable acceleration inertial loads for onboard devices produced by the rocket during launch. Two methods are commonly employed to simulate acceleration.

   i. Static load test. This test uses a hydraulic actuating cylinder and lever system to produce distributed and concentrated loads on the UUT.
   ii. Centrifuge test. The UUT is placed at one end of the long arm that rotates around a fixed axis. As the arm rotates, the UUT is subjected to inertial loads directed away from the axle.

Table 2.1 Equipment level test item statistics

Serial number	Test item	Screening	Acceptance	Routine	Qualification	Remarks
1	Insulation		✓			
2	Dielectric strength		✓			
3	Performance		✓			Under normal temperature
4	Powering bias		✓			
5	ESS	✓				Including temperature cycle and random vibration. The random vibration is tested under the fixed power spectrum density without the shock absorber
6	Aging	✓				
7	Low temperature		✓	✓		
8	High temperature		✓	✓		
9	Low-frequency scanning vibration		✓	✓		
10	High-frequency random vibration		✓	✓		
11	Noise		✓	✓		Not necessary for equipment insensitive to noise
12	Impact			✓		
13	Acceleration			✓		
14	Transport		✓	✓		
15	Life			✓		
16	Airtightness		✓			Not necessary for equipment without airtightness requirement
17	Low pressure			✓		

(continued)

Table 2.1 (continued)

Serial number	Test item	Screening	Acceptance	Routine	Qualification	Remarks
18	Humid heat			✓		When using the same UUT for multiple tests, the test order can be humid heat test—mold test—salt fog test
19	Mold				✓	
20	Salt fog				✓	Including laboratory acceleration and natural environment test. For equipment that needs to be stored for a long time in the coastal launch site, both tests are recommended. Acceleration test provides a quick reference for design, while natural environment test is more realistic, and makes up for the shortage of acceleration test which may not be fully equivalent to the actual situation
21	Thermal vacuum				✓	
22	Flight environment high temperature				✓	
23	Flight environment low temperature				✓	
24	EMC and lightning protection				✓	Including radiation sensitivity/conduction sensitivity/conduction emission/radiation emission test, each of which is based on the national standards and system requirements

2. Vibration test
   This is one of the most important tests and constitutes the largest number of experimental items in aerospace product development. Two types of vibration are considered.

   i. Random vibration. This test simulates the in-flight acoustic environment caused by engine exhaust and high-speed flight aerodynamics and transferred to onboard devices through the fairing and structure. This is a wide band random vibration.
   ii. Sinusoidal vibration. This test simulates sinusoidal and transient vibration during flight, where the vibration is caused by engine ignition, shutdown, and stage separation, with a frequency range of 5–100 Hz.

3. Shock test
   Shock environments are mainly produced by various pyrotechnic explosions, and are also called "explosion impact environments". The impact is very short and is composed of oscillations with abundant high frequencies within 20 ms. This is characterized as high acceleration, low speed, and small displacement. Generally, three methods are used to simulate shock environments:

   i. Actual explosion of pyrotechnics on real or simulated structures,
   ii. Simulation by mechanical explosive shock experimental devices, and
   iii. Simulation by shock response spectrum using a shaker.

Another important equipment level test is reliability testing, characterized as reliability growth and enhancement tests. These tests are discussed in Sect. 3.2.

2.2.2 System Level Tests

System level tests initially inherited research results from missile technologies [9], but now have significant differences from missiles with launcher's development. System level tests refer to functional and integration tests performed by integration of multiple devices and subsystems and includes system integration (SIT), simulation, etc. tests, as shown in Fig. 2.4.

All onboard avionics devices usually participate in SITs as well as the TLCS. Firing, or pyrotechnics ignition circuit and explosion tests should be performed for all newly developed launchers to examine sequential control logic, ignition current for each pyrotechnic, and also to verify if there are sneak current or loop in the ignition circuits. Firing tests are usually only performed once during the development phase.

If the rocket propulsion system is also newly developed, matching tests between the control and propulsion system will be performed on the ground, where engine ignition and shutdown functions are implemented by the control system. These tests

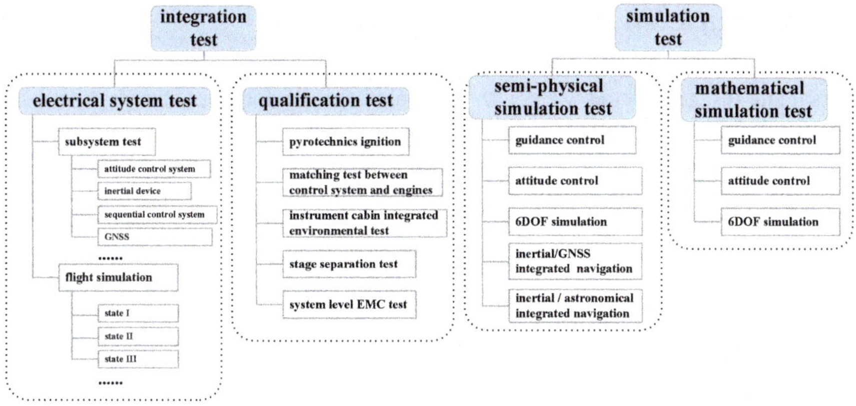

Fig. 2.4 Classification of integrated and simulation test

are closest to actual flight conditions, while the control system can be configured as required. These tests are planned separately for each new engine type and are usually only performed once during the development phase, provided they pass.

For the entire instrument cabin integrated environmental test, the electrical system is integrated into the instrument cabin (where most of the avionics devices are installed, usually at the final launcher stage) based on the flight configuration. Multiple degrees of freedom (DOF) vibrations are applied, to not only acquire cabin mechanical environment conditions but also verify the inertial measurement unit (IMU) accuracy under vibration conditions. IMU performance is comprehensively and equivalently assessed through navigation calculations.

Stage separation tests examine separation design and processes. Separation in flight is controlled by the control system, so the control system, pyrotechnics, and separation mechanism are included in separation test. It is also an effective way to check avionics device tolerances to separation impacts.

System level electromagnetic compatibility (EMC) testing is performed after each onboard equipment has passed its own EMC testing. For convenience, all UUTs are placed on testing tables, which may not be identical to onboard installation, so the test is functional. Conduction and radiation sensitivity are the main focus of the test.

Simulation tests verify guidance and attitude control algorithm design. They include mathematical and semi-physical simulations, where the real devices are combined with simulators, equivalent loads, and mathematical models. Inertial and celestial navigation, and inertial and satellite navigation are all simulated and checked. If payload characteristics are approximately the same for each launch service, semi-physical simulation only needs to be done once, and only mathematical simulation is required for each instance.

2.2.3 Launch Site Tests

Launch site test procedures vary with the rocket and payload but can be described as simply related to these major working nodes, as shown in Fig. 2.5. The work is mainly performed in two areas: the technical and launch centers (or launch zone), with most work done in the technical center, including vehicle assembly, device installation, payload integration, etc. Propellant filling and launch preparation are scheduled in the launch center.

2.2.3.1 Technical Center Activities

1. Restore and self-check the ground test and launch control equipment.
2. Unit test or check onboard devices independently. Improved reliability and maturity of avionics devices mean that most unit tests are no longer performed at the launch site. Calibrations of inertial devices, etc., remain, but these may also not be required in the future.

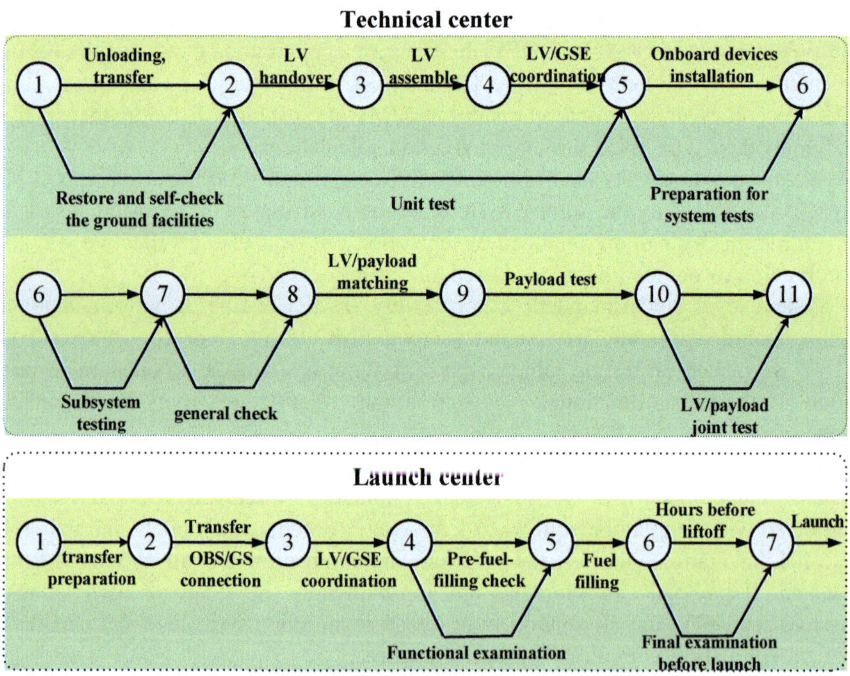

Fig. 2.5 Electrical system test process arrangement in launch site

3. Preparation for system tests. The TLCS is connected to the OBS and connections are checked.
4. Subsystem testing. The control system is divided into highly integrated subsystems, and predefined stimuli are applied to check responses. These subsystems include the inertial navigation, attitude control, sequencing control, etc., systems. Tests and response checks are performed automatically.
5. General check. Simulation of the entire flight process for the integrated system, including all prelaunch operations and launch control processes.
6. Joint LV and payload tests after they have been verified separately. Launcher test cases are the same as in the general check, and rocket and payload interfaces are also checked.

2.2.3.2 Launch Center Activities

Once the rocket and payload combination is transferred to the launch center, prelaunch checks and operations are performed.

1. Functional examination. A simplified subsystem testing, and also a comprehensive OBS self-check after the transport. Test items unsuitable if the safety concerns are omitted, such as those related to pyrotechnics.
2. Final examination before launch is scheduled after fuel filling. Key performances of vital products are verified similarly to subsystem tests.

Once the preparation activities are completed, the working process then moves to the launch control phase, and the test system is responsible for monitoring key parameters and signals.

2.2.4 *Test Technologies for Control System Software*

Software life cycle tests can be described depending on the demands of software engineering [10], as the V model (a type of software development model) that includes unit, integration, system, and acceptance testing. Acceptance tests are usually confirmed by third parties (suppliers).

There are many software testing methods [11, 12], including white, black, and gray box testing [13–15], as shown in Fig. 2.6.

Software testing research is very active, particularly for embedded software. However, few TLCS software programs are embedded, so research results on these embedded software can refer to references [16, 17], and are not discussed here in detail due to space limitations. Test technologies related to non-embedded software are presented in Sect. 6.5.6.

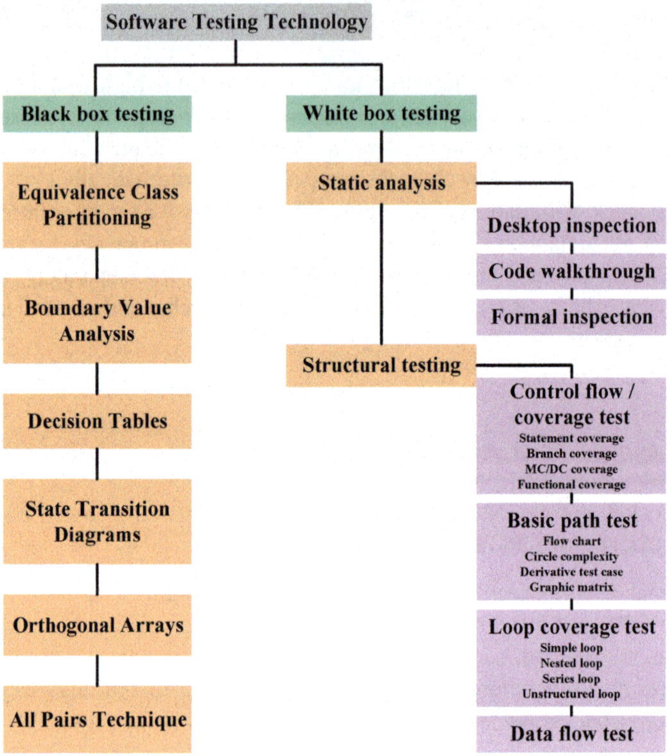

Fig. 2.6 Software testing technology

2.3 Test System Development

2.3.1 Test System Basic Design Process

2.3.1.1 Top Level Design and Requirements Analysis

Test systems (or facilities) are somewhat different from other systems, such as structure, propulsion, and control systems. While these systems can be independently developed and tested, test systems should combine each subsystem to be tested. Thus, the test system is essentially a distributed rather than centralized system. However, from the viewpoint of system engineering, test systems are still often regarded as independent subsystems, since it is convenient to allocate requirements in a traditional way and to complete system design by a professional team.

On the other hand, test systems may also be considered as a special function or ability that each system must have to ensure reliability, safety, and maintainability. With widespread promotion of built-in-test (BIT) technology, this concept has become increasingly accepted.

These two approaches reflect the diversity of understanding and demands of test systems as the test requirements become more complicated. Thus, top level test system design should be emphasized to fully understand test requirements for specific tasks, and test systems should be developed simultaneously with, rather than after, research and development of the system under test (SUT).

The test system design process follows normal system engineering concepts, such as the "SIMILAR" process [18], which can also be mapped to the classic V model. System requirements analysis, including product, process, and program requirements analysis is an important basis for test system design.

Product requirements comprise functional (FR) and non-functional (NFR) requirements. Launcher system FRs include propulsion, communication, navigation, etc., whereas system NFRs include reliability, maintainability, safety, performance, etc. Test system high-level requirements are drawn from system NFRs, and then divided into test system FRs and NFRs. Process and program requirements should also be studied. For a more detailed discussion on this, the reader can refer to [18].

After system analysis, detailed test system design can be proceeded and should also be tested, verified, and validated, similar to any other system development.

2.3.1.2 Testability Analysis

To complete the test, it is not enough to rely solely on the test system itself. The UUT/SUT must be "testable" and "easy to test", where "testable" means that there are device or system checkpoints, and the test point signals can indicate the working status; and "easy to test" means that the test operation is convenient to implement on site, particularly for the launch site and assembly plant. Analysis and design for testability plays a vital role in achieving these goals.

The testability concept originated for aircraft avionics systems, and is closely related to maintainability requirements. The definition in MIL HDBK-2165A shows that testability refers to design features that allow the system and equipment to timely and accurately determine their own working state (working, not working, performance degradation), and isolate their internal failures [19]. Design focused on improving testability is called design for test (DFT). DFT can improve fault detection coverage, shorten test time, perform hierarchically checks, and reduce maintenance costs. However, it also has some disadvantages, such as increasing hardware and software costs, and increasing design time.

Research on testability started in the middle and late 1980s in China, and national military standards were formulated, such as testability programs for material [20]. Some aerospace projects have drafted the industry standards, such as testability design criteria for aerospace products, based on engineering experience and relevant international standards [21]. However, there has not been any systematic study and application of DFT in the aerospace industry, which is mainly reflected in the following areas [22, 23].

1. Testing has been considered an auxiliary to the SUT, and has not attracted serious attention.
   Testing has been considered as only necessary in the case of failure. If the product is very reliable and production quality control is guaranteed, the test system can be simplified and the investment in it will be wasted. With upgrading quality of aerospace products in recent years, there is a growing tendency toward this view. Test equipment versatility is generally poor, with each UUT requiring a specialist measurement device, which is very costly. Inadequate DFT also makes the system being tested more complex, reducing inherent reliability, and requiring skill to maintain the balance between reliability and testability.
2. Testability requirements are qualitative, rather than quantitative.
   Although testability requirements are included in the project program, they are generally described qualitatively, with no quantitative demands for test coverage, false alarm, and fault isolation, or effective verification of these demands even if they are quantified.
3. Testability design is inconsistent with the SUT design.
   This occurs for products or systems design. For example, a relatively simple debugging tool is usually required during product development, and this gradually develops into the final test facility. However, debugging is done serially for each function, which is considerably different from the real working state. Test system development is also commonly passively performed following requirements proposed by system engineers, with little or no discussion on whether the proposed scheme is reasonable, whether test coverage is sufficient, or whether there are more optimal choices.
4. Testing methods are relatively outdated.
   Signals are generally directly sampled through wire connections, which leads to many cables, but cannot guarantee all the required information will be obtained. There is also no sensible way to determine if the result is acceptable, or to isolate failures if they are abnormal; only experienced engineers or the original designers can make the right judgments with high consequential manpower costs.

Evolved expendable LVs are only used once, and maintainability requirements are less than those of reusable LVs (RLVs), which is another factor causing little attention on testability. However, with the growth of RLVs and market competition, the demand for responsive failure detection, less on-site technical supporters, and enhanced response ability are increasing. Thus, the testing system should be systematically studied.

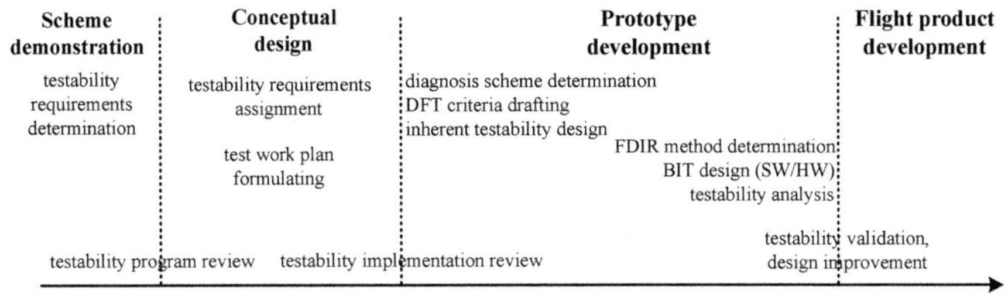

Fig. 2.7 DFT work in each development phase

Testability analysis can be scheduled in combination with launcher development in China as follows, and as shown in Fig. 2.7.

1. Test requirements should be determined during SUT demonstration.
   Test coverage and fault isolation granularity can be varied for different product life cycle scenarios. At the launch site, faults can be isolated to a device, whereas at the production factory, the fault should be isolated to specific elements or components. Automated testing unrestricted by site is preferable for some conditions, whereas manual operation can simplify the test facility for application in other conditions. These aspects should all be considered during test requirement analysis.
2. Testability requirements are assigned and work plans are drafted during SUT conceptual design.
   Test requirements are defined as part of the conceptual design requirements, i.e., not only product or system function or performance is specified but also requirements related to how to test and validate the function or performance. The test work plan should be set up simultaneously.
   In the evaluation from demonstration stage to conceptual design stage, testability program should also be reviewed.
3. Design, checking, and improvement of the test system should occur during SUT prototype development.
   Failure modes and effects analysis (FMEA) can define failure modes and causes for a specific UUT. Therefore, DFT should focus on high failure rate and risk modes, and the criteria should be specified to trade-off increasing complexity and decreasing inherent reliability. Testability analysis should be performed simultaneously, and FDI and BIT technology adopted, where appropriate.

In the evaluation from conceptual design stage to prototype development stage, testability implementation should be reviewed and confirmed.

Testability analysis and implementation should be iterative, and the trade-off carefully evaluated, where reliability is affected.

Testability validation and improvement should continue from the prototype to final flight product, as shown in Fig. 2.8.

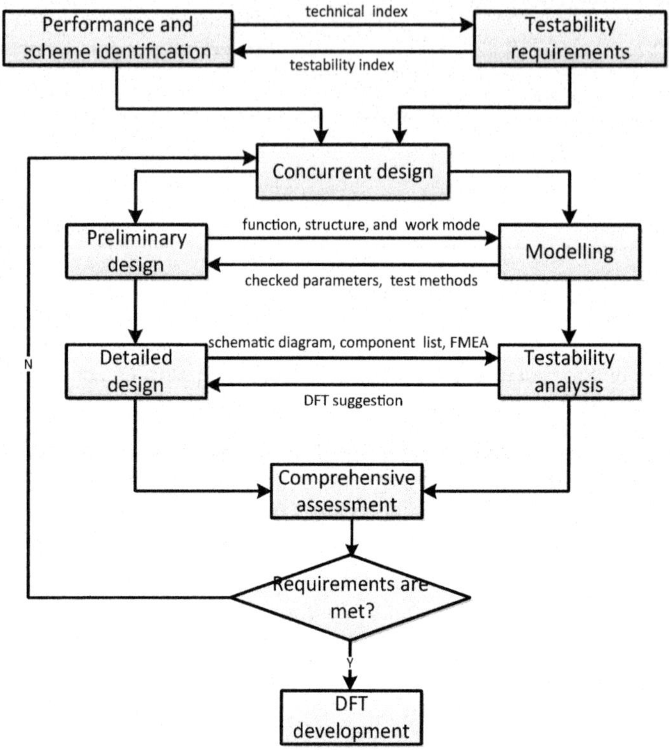

Fig. 2.8 Testability analysis and design iterative process

2.3.2 *Design for Test Approach*

Many factors should be considered in the DFT approach, as follows.

1. The test interface should be standard, generic, and simple.
2. Test points should support testing requirements at various project levels.
3. Ergonomics, automation, obstacle avoidance, accessibility, and observability should all be considered.
4. Self-inspection facility comprising BIT and sensors should be portable and easy to operate.
5. Multiple functions should be tested simultaneously where possible.
6. Parallel testing should be performed on multiple UUTs where possible.
7. System level fault detection is preferable to shorten test time as much as possible.
8. Tests should be visualized, nondestructive, and not rely on special tools.

Although many factors should be considered, there are five basic factors, as follows.

1. Control point, the direct action point where inputs can be controlled by external facilities. This is used to apply stimulation to the specific UUT state, or command the UUT to change its internal architecture.
2. Observation point, the test point where UUT status can be inspected externally to determine if the UUT is normal or not.
3. Input and output channels.
4. Fault isolation and location mechanisms.
5. Embedded BIT tests.

The following sections discuss equipment and system level DFT.

2.3.2.1 Equipment Level Design for Test

Equipment level tests are usually performed in design departments or production factories, which provides better testing resources, and are not limited by the site. There are several ways to complete the tests.

1. Key signals sent to the special connector.
   The specific connector(s) is(are) reserved to test key signals and verify UUT performance. During flight, the connector(s) may be shielded.
2. Boundary scan based BIT test.
   Boundary scans (IEEE1149.1) were first used to test digital integrated circuits, and were gradually adopted for analog circuits. The test requires boundary scan components, which have specific internal architecture, including test access ports (TAPs), TAP controller, and instruction and data registers. All components without time limits can be connected in a standard static state and tested. Boundary scan units are generally located between the core logic and periphery pins, to check resistance, logic gates, IC drive, RAM, FLASH, crystal oscillators, etc.
   However, boundary scan components are somewhat specialized, and their application is limited.
3. Processor-controlled BIT test.
   If a processor is integrated in the UUT, then BIT tests can be controlled by the processor while the main work is performed, if there is sufficient spare running time. Using A/D, V/F, I/F, and electronic switch components, the input and output signals, voltage, and running status (e.g., register status) can be sampled and forwarded to external terminals for analysis and verification. Many self-data-sampling circuits are integrated and the method has become widely applied.

2.3.2.2 System Level Design for Test

The basic system level DFT principles are as follows.

1. A complex system should be divided into adequate modules to simplify testing.
2. Test functions are integrated in the SUT. The module is checked first, then the interaction between modules, and then the whole system. For the launcher, this can be achieved as follows.

    a. Special connectors for system level testing.
    Vital device signals are sent to the connectors, called DC connectors in China, and are sampled directly by the ground test equipment. However, DC connector limitations mean only a few signals can be checked. In addition, many signals are not suitable for sampling through long wires, so the DC connectors are generally set in the instrument cabin or upper stage. These must be disconnected before liftoff to facilitate prelaunch operations, hence these signals cannot be monitored before and during takeoff.
    With increasing demand to simplify launch zone operations, this method has been rarely adopted in new launchers.
    b. OBC-based testing.
    OBC-based testing can help avoid DC connector limitations. For example, IMU output signals can be directly sampled by OBC and relayed to the ground facilities. OBC can also transmit commands itself to check onboard attitude control systems independent of ground test facilities.
    c. Hierarchical integration boundary scan technology.
    Boundary scan technology can be extended to hierarchical integration testing, with the boards or modules interconnected through IEEE1149.1 standards, and the devices and system integrated through the IEEE1149.5TM bus. The whole system can be connected with the same hierarchical structure, including system, subsystem, equipment, boards, etc. Testability mechanisms of the different levels are connected by the test bus.
    d. Testing using the telemetry system.
    If signals are handled by the telemetry system, the ground test system can share the data, particularly for those that fluctuate slowly, or are less real-time, or less important.
    e. Test by each unit itself (BIT).
    If the UUT is digital and relatively intelligent, and it can implement BIT employing its own processing capacity. It can also sample signals of nearby units that have relative low process capacity. The sampled data can be relayed to the GS through high-speed communication channels and analyzed on ground terminals.
    BIT technology can dramatically simplify connections between the UUT/SUT and test system, while providing more information. Therefore, fault diagnosis expert systems for health monitoring and management have emerged, which are discussed in Chap. 7.

The following requirements shall be considered for testing at the launch site.

1. Ensure product and test operator safety.
   For example, after pyrotechnics installation, great care should be taken when inspecting ignition and sequential control circuits.
2. Ensure UUT/SUT installation and connection status remains the same as in flight during testing.
   For example, do not disconnect a connector to check something and then restore it after testing. This could significantly increase the site workload and cause unreliable effects.
3. Test depending on the flight profile, i.e., *test what you fly*.
   For example, ensure the closed-loop profile simulates the flight process as much as possible.

2.3.3 Standard Buses-Based Test System

Open and standard architectures provide testing convenience through the adoption of automation and standard buses. Almost all vendors emphasize product openness and compatibility, reducing the investment required for testing, and users can replace modules or software to meet various scenario requirements. These features promote application of these technologies and products in LV TLCS [24, 25].

This section discusses current widely applied bus technologies, and also introduces several new bus technologies either on the horizon, or recently applied in TLCS.

2.3.3.1 Common Bus Test Technologies

GPIB

As early as the 1970s, IEEE defined a set of high-speed data transmission protocols (488.1/488.2) to control many independent measurement instruments from a single computer, and general-purpose interface bus (GPIB) technology was the early development stage of the virtual instrument in this standard. Thus, GPIB facilitates changing test activities from stand-alone manual operation to large-scale automatic testing.

Test systems comprise a personal computer (PC), GPIB interface card, and instrument, connected by GPIB cables using a common string grammar (488.2 or SCPI) to control the instrument. The GPIB interface can support as many as 14 instruments, with cable lengths up to 40 m, connecting many instruments to form a system in any convenient topology (star, serial, etc.). However, GPIB cannot provide synchronized triggering for many instruments, and the bandwidth is insufficient for transmitting huge data volume. Therefore, GPIB is mainly used for desktop instruments, and for high accuracy and low transmission rate applications.

VXI/PXI

The VME extensions for instrumentation (VXI) and VME bus switched serial (VXS) data buses developed from the VME bus, and cPCI, PXI, PCI-Express, etc., developed from the PCI bus. The VXI bus is regarded as the first open, international and test-oriented instrument bus, and has been applied in the Chinese manned spaceflight project [26]. It is an extension of the VME bus that incorporates system reference clock, trigger bus, timing, and synchronization functions required for sophisticated testing. The bus has stable power supply, powerful cooling ability, and strict RF/electromagnetic interference (RFI/EMI) shielding, and has been widely used.

The PCI extensions for instrumentation (PXI) bus [27] adds some mature technical specifications and requirements on the PCI bus kernel, and multi-board synchronization trigger bus specifications, to meet high precision timing, synchronization, and data communication requirements. It also refers to the VXI standard.

Initially, PXI could only use embedded controllers, but it has been extended to outer PCs connected through an adapter in the chassis. These adoptions allow direct PC control, chassis expansion, and longer distance control. Figure 2.9 shows the extension of a PXI system using the Starfabric serial bus adapter, and Fig. 2.10 shows the adapter diagram for this configuration.

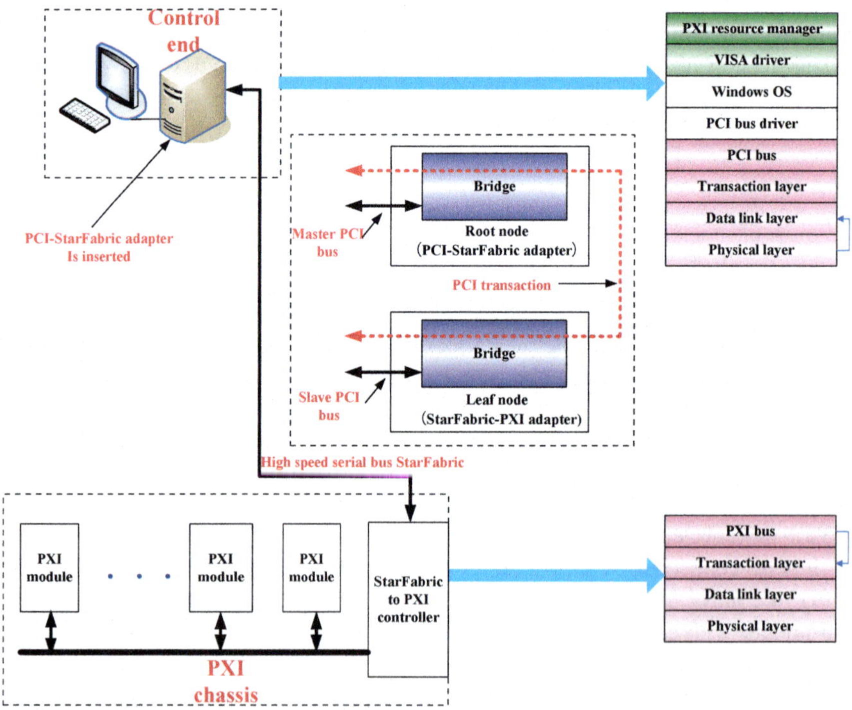

Fig. 2.9 PXI adapter system block diagram

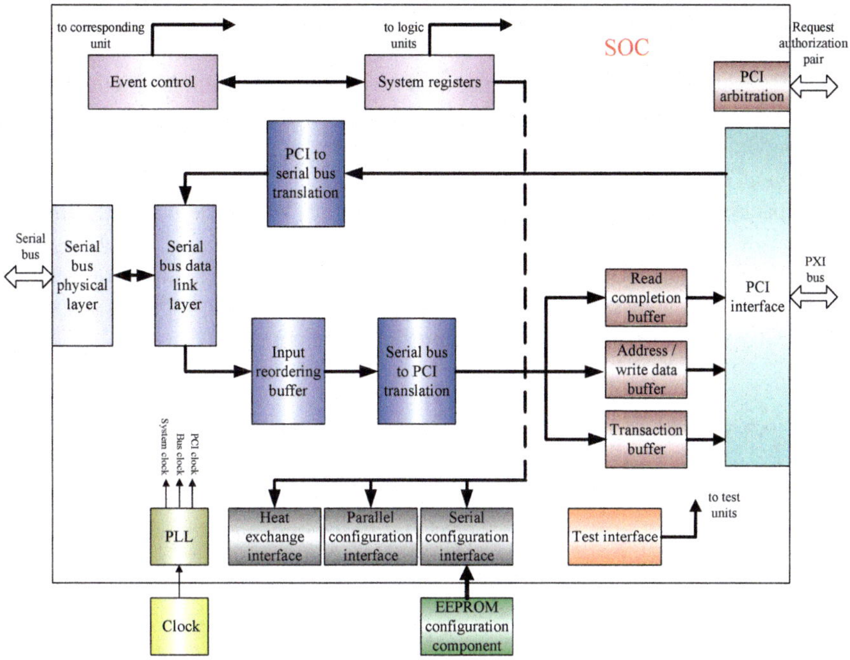

Fig. 2.10 Starfabric-PXI adapter hardware scheme

For Starfabric serial buses, the PCI-Starfabric adapter and Starfabric-PXI controller implement peer-to-peer functionality, including physical, data link and transaction layer, and PCI/PXI interface logic. The physical layer completes the serial-to-parallel conversion of the low-voltage differential signal (LVDS) signal encoded by the high-speed 8b/10b bus. The byte stream is then packed, and its integrity checked by the data link layer. The data packet comprises a sequence number, content, and CRC check; if a packet passes the integrity check, the package will be received, otherwise it will be discarded. Received packets are processed into the transaction layer, which translates data from the Starfabric bus. The translated signal then appears on the PCI bus, and if it is a trigger signal, it will be forwarded to the PXI trigger bus.

Hybrid Systems

One of the advantages of GPIB is that it can connect different test equipment together to form a system. VXI and PXI bus compositions are more concise, with all devices integrated in one case. However, not all test equipment can be produced as cards. Thus, the LXI bus was developed to incorporate both designs, connecting various chassis by LAN networking rather than GPIB.

A complex test system is rarely constructed using only stand-alone instruments or chassis-style equipment, and these will combine into more configurations in the future. By not considering inheritance and strict environmental restrictions, low cost, low power, and physically smaller bus systems will be even more acceptable. Single module functionality will also continue to increase, promoting competition for different bus systems. For example, PLC, which is also the architecture with chassis-backplane-module, may incorporate PXI and VXI system functionalities. Serial buses are fundamentally more efficient than parallel buses for high-speed communication. Therefore, future test systems must allow coexistence of parallel and serial buses, high-speed networks, etc., and it is also possible to construct TLCS using serial bus and network systems simultaneously.

2.3.3.2 VPX Bus

System designers have gradually recognized the limitations of shared parallel buses, moving to high-speed serial buses. These latter buses implement serial transmission protocols with low-voltage differential signals using a clock package, clock recovery techniques, and signal pre-emphasis and equalization, to achieve serial transmission rates up to 10 Gbps in a single pair line.

VPX is a new generation high-speed serial bus [28, 29], intended to protect and continue the VME bus applications, but incorporating many innovations. Aside from some retained VME signal definitions, the VPX definition of electrical signal is considerably different to the VME standard. However, the mechanical structure is similar, retaining the advantages of cold conduction and vibration absorption, meeting the reinforced demands of embedded systems.

These specifications have been applied to the VITA46 series, with all connectors replaced by MultiGigRT2 connectors to support high-speed differential signal transmission and more signal pins, as shown in Fig. 2.11. The connector is a silicon wafer structure with tightly connected contacts and internal wiring of microprinted circuit boards, and has low insertion loss, and low bit error rates. The silicon wafers incorporate electrostatic discharge ground and contact layers to prevent accidental discharging during operation.

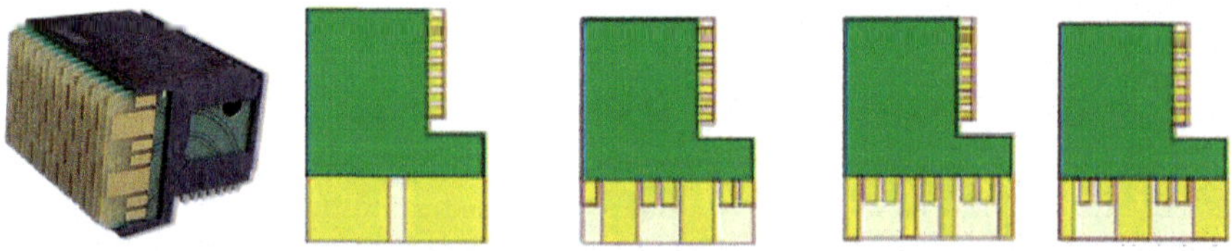

(a) Power signal (b) Single end and differential mixed signal (c) Single end signal (d) Differential signal

Fig. 2.11 MultiGigRT2 connectors

There are a variety of high-performance switching architectures available, with gigabit Ethernet (GbE), serial rapid IO (SRIO), and PCI-Express (PCI-E) particularly prominent.

- GbE is an IP-based data communication standard, and can be adopted by networks between platforms or subsystems in the backplanes. The technology is suitable for loosely coupled system links, with Ethernet being the most common.
- SRIO is the best method for high-density cluster interconnection of multiprocessors for digital signal processing applications. It is more suitable for communication between a large number of processors than PCI-E or Ethernet.
- PCI-E is a better choice for tightly coupled communication between processors, peripherals, and application boards. It is widely used in PCs, and has the advantage of low cost.

VPX board signals are logically divided into the following layers.

1. Application layer, including power pins and various application signals.
2. Management layer, or intelligent platform management bus, including onboard intelligent platform management controller and chassis management controller for fault prediction and diagnosis. This provides low power sources, defined by VITA46.0 and 46.11.
3. Control layer, for reliable communication between various application control and development, typically using a gigabit Ethernet interface.
4. Data layer, enabling high throughput data exchange, typically using PCI-E and SRIO interfaces.
5. Extended layer, mainly used for I/O interfaces, VME bus bridging, etc.

Using the serial bus, VPX cards can form a variety of flexible topologies, generally divided into four types [30, 31], as follows.

1. Exchange. All boards communicate with each other through switches (boards) and dual redundancy can be adopted to improve reliability.
2. Master–slave. Similar to traditional computer architecture with the master board installed in the root slot controlling slave boards installed in leaf slots.
3. Whole network. All boards are connected point to point, which has high reliability and any connection failure does not affect other boards.
4. Hybrid. VPX and VME interfaces are integrated and compatible.

2.3.3.3 Real-Time Ethernet

Real-time Ethernet has the advantage of ethernet flexibility, which provides access convenience and maintains high bandwidth and speed. It also provides strong real-time performance for control applications. Time-triggered Ethernet (TTE) and

Ethernet for Plant Automation are more mature, and we restrict the discussion to TTE in this section.

TTE provides time triggering using standard Ethernet protocols, with a global unified clock established in the network system, on which communication between the terminals is based upon. Therefore, TTE is a deterministic integrated Ethernet with synchronous and asynchronous communication functions, and is mainly applicable to the following situations.

1. Systems with predominantly periodic information (periodic uplink control instructions, data inputs, downlink digital telemetry, engineering parameters, scientific data, etc.) rather than nonperiodic (random uplink and downlink data, burst data, etc.).
2. Huge volume data transmission is usually short-term and occurs in specific time phases.
3. The time accuracy of the control channel is very high to the nanosecond level, and the backbone network time accuracy of 100 μs is enough for the applications.
4. Size and the number of backbone network nodes can be planned in the initial system design stage, and predicted so network load is relatively stable.

TTE has strong fault tolerance, with dual and triple redundancy support. Network architecture is also very flexible, supporting star, line, tree, and ring topologies. Communication rate can reach 100 Mb/s or 1 Gb/s for copper wire or optical fiber, respectively.

TTE communication protocols include three types.

1. Time-triggered or hard real-time communication

Each terminal and switch begins transmitting and receiving data at specified times, depending on the configurations of the message descriptor list (MEDL). The whole network MEDL is generated off-line by special tools, so shared resources conflicts are avoided. Each terminal in the network transmits at most one message at a time, ensuring no competition for the output link, and all network operations are triggered based on global time. In the TTE protocol, "real-time" refers to a transmission delay of no more than 100 μs, and precision error is not more than 1 μs.

The TTE messages are arranged by cluster cycle depending on the minimum common multiple of different information cycles. Figure 2.12 shows the

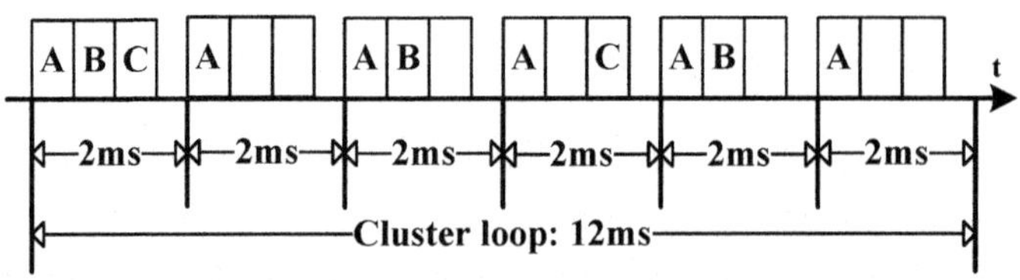

Fig. 2.12 Communication arrangement of different cycle information

communication arrangement for applications A, B, and C with 2, 4, and 6 ms cycles, respectively.

2. Rate-constrained or soft real-time communication

When there is no time-triggered communication in a network, rate-constrained communication can be performed depending on a predefined bandwidth. It is used primarily for less stringent real-time requirements than time-triggered communications, where occasionally delays are tolerated. These data are typically not sent periodically over a fixed time interval but are sent when required, with a portion of bandwidth reserved to ensure these services.

3. Standard Ethernet or best-effort non-real-time communication

Standard Ethernet communication is commonly described as best-effort. It is event triggered, and when none of the above two data packets are at the switches, common data messages are sent. If there is time-triggered or rate-constrained communication, standard Ethernet communications are interrupted. A suitable application could be transmitting diagnosis data that has no strict time requirements, and the faults are not critical.

Three types of protocols can be classified as two information types: time-triggered (TT), for time-triggered communication; and event-triggered (ET), including rate-constrained (RC) and best-effort (BE) communication. The TTE switch's main function is to identify the information types. Since TT data is sent at preemptively fixed time intervals, when TT data needs to be transmitted during the ET process, ET information is stored in the buffer, and restarted automatically once the TT data is sent. If ET data needs to be transmitted during the TT process, it can also be stored in the buffer and sent at the end of the TT transmission.

When TT and RC messages are transmitted simultaneously, TT information has higher priority, delaying RC transmission. When there is no TT information, RC is prioritized, and BE has the lowest priority. Figure 2.13 illustrates transmission time relationships between the three signals.

Figure 2.14 shows information processing in the protocol stack. Priority of the real-time channel is higher than that of the standard ethernet channel. The real-time scheduling layer is added on the MAC, to schedule ordinary and real-time data. Preemptive scheduling based on priority for these data is adopted in the real-time scheduling layer.

The TTE adopts a network topology, and clusters can be connected by a group of terminals through redundant or multiple channels and switches, with different channels on different switches to provide high reliability communication. Large

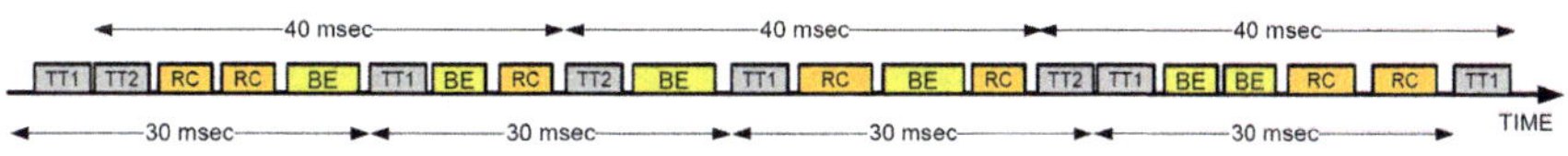

Fig. 2.13 Communication arrangement of different priority information

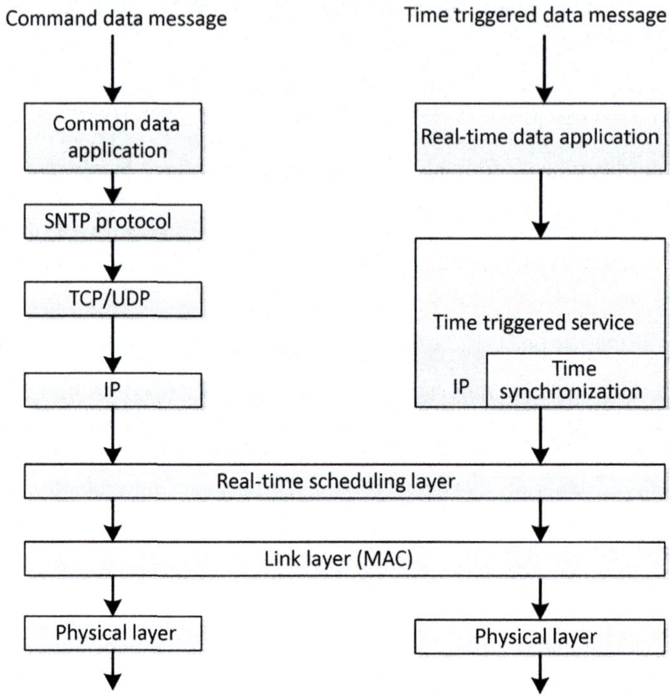

Fig. 2.14 TTE protocol stack information processing function diagram

communication networks can be constructed with typical network topology as shown in Fig. 2.15.

Traditional Ethernet terminals can be connected to TTE switches or standard Ethernet switches, but TTE terminals can only be connected to TTE switches. The host connects to the TTE controller via the controller network interface (CNI).

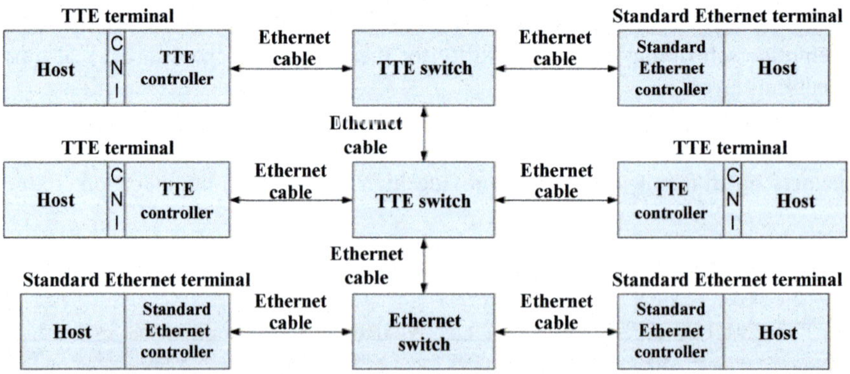

Fig. 2.15 Typical TTE network topology

For more detailed discussion on TTE technologies, the reader can refer to the relevant references [32–37].

2.3.4 Natural Environment Adaptability Design

Ground test equipment is subjected to many natural environmental impacts compared with OBSs, particularly at coastal launch sites. Therefore, this section describes natural environment adaptability design for ground equipment. For other adaptive design details, such as force, temperature, and other environmental impacts, the reader can refer to the OBS design guidelines.

Enclosed moisture proofing, salt fog proofing, and fungus-proof ("triple proof" in short) chassis can provide relatively favorable conditions for test facility compared with poor external environments. The chassis comprises structural parts, a power supply, backplane bus, heat exchanger, etc. The heat exchanger isolates the inside and outside air, providing a good working environment. Internal protection should reach IP65 to completely prevent invasion of foreign objects, dust, and water splash, which can damage electrical devices.

2.3.4.1 Enclosed Chassis Structural Design

The enclosed chassis comprises a main shell, heat exchanger, ventilation, connectors, etc., with typical layout as shown in Fig. 2.16.

The outer shell is made of aluminum alloy with conductive oxidation, overlaying organic coating on the outer side to meet triple proof requirements. The connection between the shell and the cover are sealed with a sealing ring embedded in a

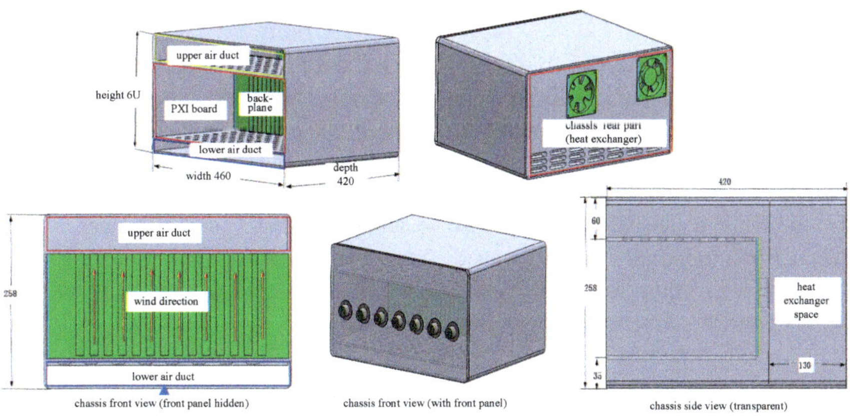

Fig. 2.16 Enclosed "triple proof" chassis

Fig. 2.17 Gore ventilation device

groove. A gas permeable device is attached to the side of the chassis, so water vapor can be discharged and air can be ventilated at the same time. The ventilation device balances the pressure inside and outside the chassis due to temperature changes, prevents pressure build-up, and consequential shell or seal damage, while also preventing moisture and salt entry into the chassis. Figure 2.17 shows an example application of the Gore ventilation device.

The front panel of the chassis incorporates a conductive film shielded window, allowing monitoring of the working state inside the enclosure while satisfying EMC requirements.

2.3.4.2 Chassis Heat Dissipation

In operation, input power is converted into heat and causes inside temperatures to increase. This will increase component failure rate, and reduce equipment reliability. Hence, some measures are required to cool the components. Thermal design is an effective way to improve equipment reliability, and usually the temperature inside the chassis is controlled within a certain range, e.g., 0–55 °C.

Cooling the enclosed chassis can also be achieved using a heat exchanger, which isolates the internal air from the outside environment, as shown in Fig. 2.18. The heat exchanger comprises a shell, copper foil heat exchange core, internal and external fan, air duct separator panel, heat exchanger controller, etc. The main parameters affecting heat exchange performance are the air volume and wind pressure of the inside and outside fan, number of copper foils, copper foil structure, etc. For coastal launch sites, the external fan should be salt fog proof.

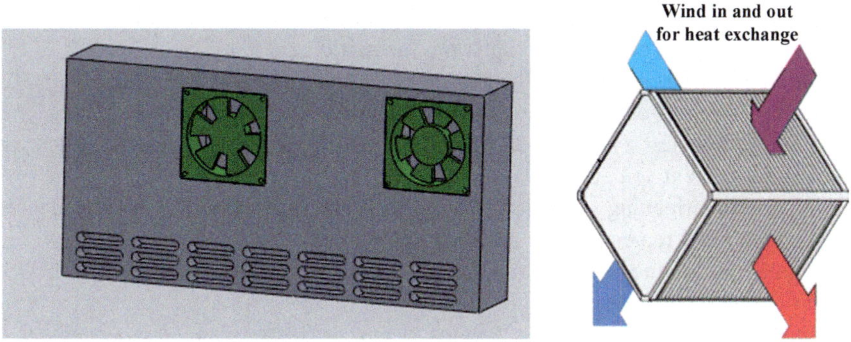

Fig. 2.18 Heat exchanger and exchange core

The heat exchanger core comprises multilayer copper foil, with a few mm between the foils, as shown in Fig. 2.18. Internal hot air flows through the heat exchanger core, and exchanges heat with the cold air on the other side of the copper foil. The cooled air then flows back into the chassis, while the now-warmed external air is expelled to the outside. The two airflows never mix, being separated by the copper foils, ensuring cleanliness inside the chassis, as shown in Fig. 2.19.

2.3.5 Test System Functional Verification

As discussed in Sect. 2.3.1, the test system must also be checked, so a corresponding operating environment should be constructed, called an equivalent device, which simulates the UUT/SUT receiving stimulus or control commands, and returns simulated feedback to the test system. Typical equivalent device features for TLCS include the following.

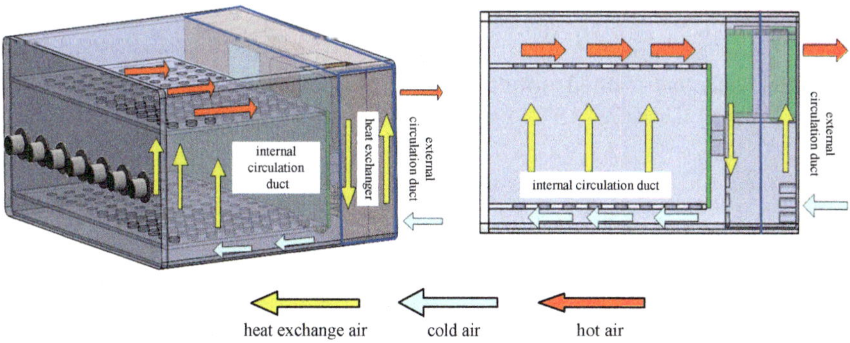

Fig. 2.19 Heat dissipating duct of the chassis

1. Provide a response to control commands sent by the GS, such as feedback of power on or off, or replies to software instructions.
2. Simulate various signals that can be measured by the test system, including AC and DC voltage, current, sequence, frequency, temperature, etc.
3. Simulate real test system loads, and check the test system capacity, such as constant current source capacity, and ignition control load, etc.
4. Simulate interlocking conditions of launch control processes, mainly feedback from the OBS representing current status.
5. Simulate OBS and GS communication, particularly OBC responses to TLCS and data transmission.
6. Simulate OBS outputs when failures occur, and check the test system can identify the different failures and report them in appropriate time.

The equivalent device should also simulate other subsystems when needed. For example, when the control system is checked alone, the equivalent device should simulate all the interfaces required for the control system, including telemetry equipment, sensors, flight abort system, etc.

Equivalent device design can refer to standard bus-based measurement systems, and should have a good user interface. Upper and lower computers should be configured with a real-time operating system or high-performance embedded devices applied in the lower computer for sampling and command execution to meet real-time requirements. A commercial operating system should be installed in the upper computer to set up test profiles, control the test process and display record, and retrieve data.

2.4 Test Technology Development

To enhance LV competitiveness, space transportation systems must become more responsive, reliable, and economical. Therefore, test scheme must enhance virtual verification in early development phases, improve test credibility and coverage, reduce operation state differences between testing on ground and flight in space, and improve overall testing efficiency.

The following section focuses on system level virtual tests, integrated environmental tests, and methods to reduce costs. Responsive test and launch control are discussed in Chap. 7.

2.4.1 System Level Virtual Test

When there is no collaborative design and virtual test platform, design and verification should be carried out separately and in serial, and the design can only be checked when the physical prototype is complete. However, virtual test technology [38] can set up an integrated, optimized virtual environment for design and verification, enhancing verification ability of early designs.

For example, a closed loop feedback control system can comprise a virtual system before any device is manufactured, and simulation verification can be performed with the required tools. The virtual elements are usually divided into the following.

1. Digital circuits;
2. Analog/mixed signals;
3. Embedded software; and
4. Sensors, actuators, other nonelectrical products, etc.

Joint simulation of digital, analog and mixed signals can be achieved with support of various commercial simulation tools or software programs. Sensors, actuators, and other nonelectrical devices are commonly represented by mathematical models (such as inertial device models), equivalent circuit models, or transfer functions (such as servo mechanisms). The IEEE VHDL-AMS (hardware description language, and analog and mixed-signal extensions) language is widely used in mixed signal modeling in many fields. Depending on the behavior or performance of the object, it can construct a virtual system including avionics, hydraulic equipment, and motors, where the thermal and magnetic analysis can be performed. Current modeling and simulation technologies make it feasible for system level virtual verification and testing.

Virtual prototypes are the basis and key of simulation, and their design and simulation can be performed hierarchically. Lower design layers include finer model descriptions, whereas the top level model only considers the function and important performance metrics. Figure 2.20 shows an example of an integrated hardware/software integrated simulation environment [39].

The CPU models can be constructed in a variety of ways:

1. Complete processor model (register transfer level), written in VHDL;
2. Simplified processor model (transaction level); and
3. Behavioral model written in a high-level language,

where the more complete the model, the slower the simulation speed, and the lower the efficiency.

Application software (including flight control software, embedded operating systems, etc.) is compiled into the CPU target code using an instruction set simulator (ISS). ISS is a virtual microprocessor that decodes and executes the target code, and interacts with the hardware emulator externally through the processor bus function model (BFM). BFM completes the conversion from instruction to cycle level, produces bus cycle sequences, realizes the bus interface, and drives these signals to the hardware simulation environment. It also samples bus cycle responses and transmits them to the software environment.

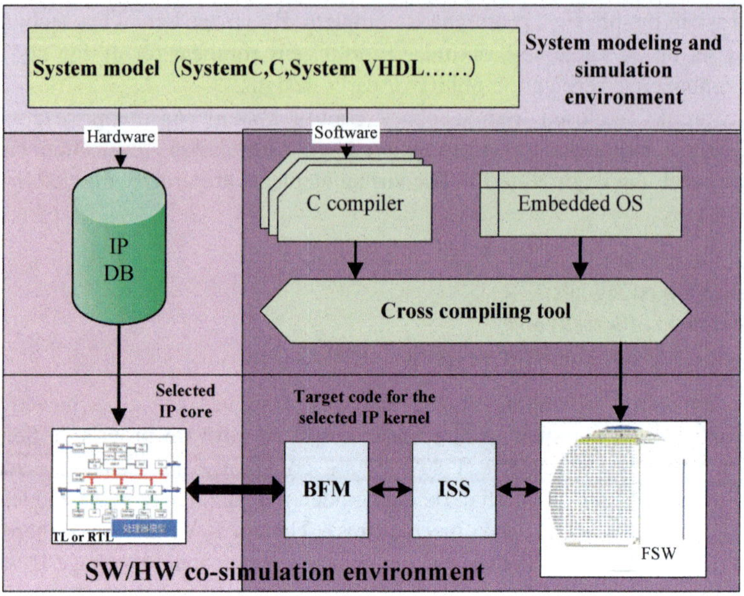

Fig. 2.20 Software and hardware integrated simulation design and verification environment

With a virtual prototype in the conceptual design phase supported from simulation tools, system function and performance can be examined quantitatively. Thus, difficult or costly experiments, such as boundary tests, worst case analysis, margin limits tests, etc., can be conveniently performed.

Nowadays, virtual test technology has been extended from avionics systems to structures [40] and propulsion systems, e.g., virtual vibration tests [41–44], etc. Consider Japanese new generation flagship LV H-X as an example.

When Japan developed the H-X, one of its outstanding features was a more robust and reliable engine, and modeling and analysis [45, 46] was the most effective way to achieve this goal. First, all the failure modes were identified based on experience and fault tree (or event tree) analysis. Then, model-based quantitative risk assessment was adopted. Probabilistic design analysis was used to assess component-level test results, and also for analytic model analysis based on failure mechanisms. System reliability was derived from the fault tree, and design improvements and component level testing plans were drafted. Thus, the analytical model validity was evaluated through quantitative analysis of uncertainty. This was a major departure from previous attempts, which relied heavily on physical engine tests, with considerably increased costs.

A virtual simulation environment is adopted for the H-X's LE-X engine. This environment included two parts: failure mode identification and redesign, and model-based evaluation and validation. The activities of model-based evaluation of reliability, risk assessment, sensitivity analysis, and reliability control are common aspects of the two parts. Failure mode identification was based on design results and a KB, which were input for model-based reliability evaluation, creating a closed loop redesign process. The KB was composed of failure information from which lessons were learned. Model-based evaluation of reliability was performed on the basis of analytical model improvements, and elemental test results, creating a closed loop for uncertainty reduction. Reliability control was to improve the design, and also reduce uncertainty, its final outputs were applied to product characteristic control for screening, while test-validated uncertainties were used to update reliability.

Figure 2.21 shows a relatively complete virtual test system framework for the control system based on the various virtual test technologies, e.g., virtual noise [47] and thermal [48] tests. Thus, this framework can be considered a multidisciplinary simulation test platform.

The virtual test results can be quickly returned to the design department, allowing verification to move forward and avoid repeated design alterations. Alterations in this phase are extremely cost-effective, since there was no investment in physical products.

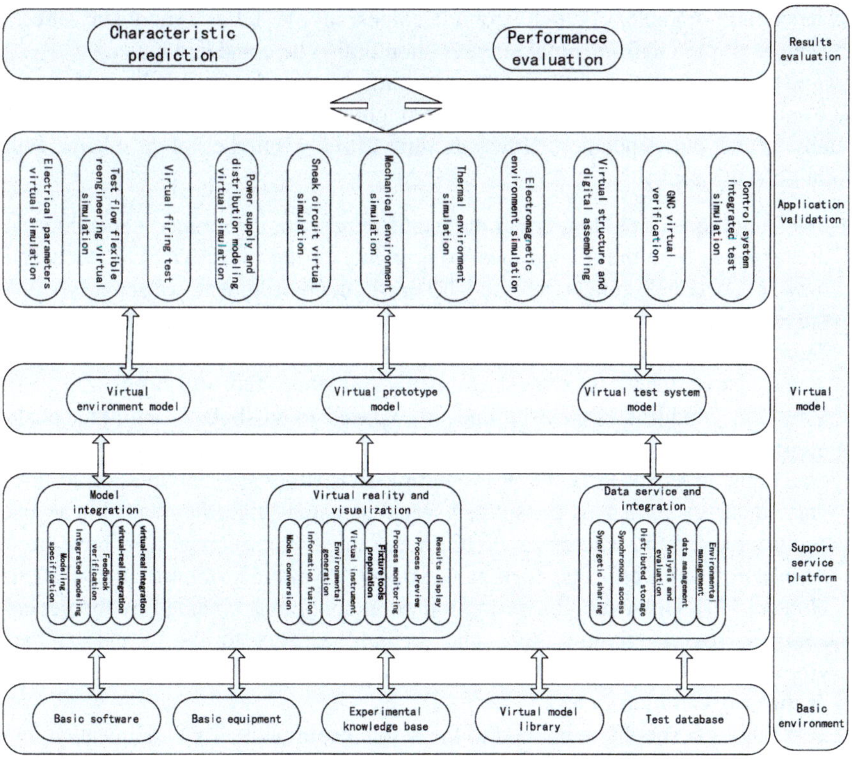

Fig. 2.21 Technical framework for the control system virtual test

2.4.2 System Level Integrated Environmental Test

2.4.2.1 Environmental Conditions Synthesis

Aerospace products will experience various environmental stresses during flight [5, 8]. However, environmental tests are usually performed individually, due to costs, schedules, and for convenience. Understanding of the difference between ground test and flight conditions has deepened in recent years, and modern tests reconstruct real flight environments as much as possible. If there are a variety of environment stresses, we should also consider their simultaneous effects.

The US air force has long known that MTBFs evaluated in the laboratory are much longer than on-site outcomes, with a typical ratio of 6.5:1, respectively, due to testing condition differences. Thus, the US Air Force CERT plan requires that not only all environmental stresses, such as vibration, temperature, altitude, and humidity be applied to all UUTs but also requires the stresses be made over some time to simulate practical mission profiles. The combination of acceleration and vibration stresses will have adverse effects on rocket structural parts, hence comprehensive environmental testing facilities—including centrifuges and vibration tables—have been constructed in the US and France. With the application of more advanced and complex optoelectronic devices in the LVs, synergistic effects between multiple environmental stresses should also be considered.

Consider the mechanical environment test as an example. Generally, rockets are huge and the installed device is scattered throughout the rocket. Hence, the device is usually tested independently. However, this raises a number of problems with vibration test validity.

- Test conditions, including environmental stresses, are determined depending on the response from the limited measuring points during rocket vibration tests and may not accurately reflect the true UUT in-flight environment, particularly when the measuring point is far from the UUT. Therefore, test conditions are always determined conservatively, to reduce risk, which increases OBS cost and weight. Nevertheless, overstress or insufficient stress may still occur.
- Vibration conditions that a product is exposed to in flight is simultaneously multiple directions. However, the vibration tests are performed in three axial directions, independently. Even in multiple axis simultaneous vibration testing, only linear movement is considered, whereas actual flight has both linear and angular movement. Therefore, 6-DOF mechanical environment tests have been applied for some projects, such as the 6-DOF hydraulic vibrator at ESTEC in Holland [49]. However, converting vibration conditions obtained by theoretical analysis, rocket vibration test, and in-flight sensors to linear and angular vibration test conditions is difficult, particularly when assigning the proportions of the two vibration types.
- Test vibration should be applied to the whole cabin with all UUTs installed just as they would be in flight, to simulate practical situations.

- Acceleration response is the most common vibration control parameter, which can also cause problems. Therefore, force limiting technology [50], which limits vibrational force applied to the UUT, has been applied for multi-axis vibration tests to more reasonably examine the product.

Similar problems exist for thermal environment tests. Heat radiation conditions are often converted to high-temperature testing, but this conversion fails to consider reflection, sheltering, and other characteristics of the temperature field, raising the likelihood of inappropriate stresses applied during testing.

Table 2.2 shows the influences of different environmental combinations on device. Combinations of influences that reinforce each other should be considered and applied in the test as much as possible.

Table 2.2 Influence of environmental combinations on device

Serial number	Environmental combinations	Influence
1	High temperature and humidity	High temperature increases the permeability of humidity. The damage effect of humidity is generally serious with the increase of temperature
2	High temperature and low pressure	When the pressure drops, the material outgassing increases; and if the temperature increases, the material outgassing increases too; therefore, the two are mutually reinforcing
3	High temperature, shock, and vibration	The effects on the properties of the materials will be strengthened, and the degree of influence is related to the environmental intensity. The polymer is most vulnerable to this environmental combination effect
4	High temperature and acceleration	Similar to No. 3
5	Low temperature and humidity	With the temperature decrease, the relative humidity will increase, which leads to the structure strength decrease and the electrical property change. Low temperature can also cause moisture condensation
6	Low temperature and low pressure	The combination accelerates the seal leakage
7	Low temperature, shock, and vibration	Low temperature can strengthen the effect of shock and vibration
8	Low temperature and acceleration	Similar to No. 7
9	Humidity and low pressure	Humidity increases the effect of low pressure, especially for electronic device. The actual effect is mainly determined by the temperature
10	Humidity and vibration	The combination speeds up the breakdown of electrical materials
11	Low pressure and vibration	In particular, the combination will have an enhanced effect on electronic device

2.4.2.2 Unit Under Test Integration

Whole cabin tests, such as the instrument cabin test, have the advantage that the electromagnetic environment is more real. When a single UUT is tested, EMC cannot be effectively verified since the devices are laid randomly on a table. Whereas in the cabin, connections between UUTs, and the electromagnetic environment are very close to actual flight conditions.

2.4.2.3 Test State Equivalence

Traditional equipment level environment tests focus on function, but performance will be affected under different environmental conditions, and the UUT function will fail when the performance degrades sufficiently. Therefore, performance testing should be considered concurrently in environmental testing.

For example, IMU calibration cannot be performed during environmental tests. Therefore, the IMU is checked before and after testing under a static state with only basic functional monitoring during the environmental test. However, the IMU is often in dynamic and vibration overload during flight and IMU dynamic accuracy has been shown to be significantly different from the static calibration value. Thus, evaluating IMU dynamic accuracy is important.

In summary, integrated test environment should be further studied, to more effectively create electrical, magnetic, thermal, and mechanical conditions for UUT/SUT, and establish a comprehensive environmental test platform, while also considering cost and feasibility. Installation, interfaces, and working profiles should be simulated as close as possible to actual flight status. Performance, not just function tests, should be implemented to identify environmental influences, and catch fault symptoms in advance.

2.4.3 Test Technology for Expense Cutting

With increasing reliability requirements, a great deal of time, manpower, and revenue are consumed in testing, which inevitably reduces the rocket's competitiveness. New test methods and approaches, based on product failure mechanism studies, can quickly meet user requirements while reducing these costs. There are a number of studies and considerations, as discussed in the following sections.

2.4.3.1 Shortening Test Time

The high accelerated life (HALT) and high accelerated stress screening (HASS) tests offer significantly shorter tests. HALT uses the accelerated stress for life test without changing the failure mechanism, hence shortening test time while improving test efficiency and reducing cost [51, 52]. The method is more effective if HALT and practical failure modes are the same. However, the accelerated stress conditions may not occur in practical applications. Therefore, many experiments are required to determine the acceleration model and suitable acceleration factors.

The HASS test uses optimized screening without affecting product reliability. HASS applies extreme nondestructive temperature stress and temperature change rate. The most extreme conditions are generally determined in HALT, while HASS conditions are slightly less, i.e., HALT is the basis for HASS. HASS has been successfully applied for screening inertial devices [53].

2.4.3.2 Reliability Testing and Analysis with Extremely Small Samples

Reliability tests require a certain number of samples to assess performance. However, sample sizes are often small or very small for launchers, with commonly only 1–2 products available for reliability tests, to save investment on testing samples. Therefore, reliability assessments based on small sample sizes have been the focus of the research, as is storage reliability prediction determined by periodic inspection, or product life assessment not yet having a generic processing method.

2.4.3.3 COTS Component Upgrade Screening

Previous test techniques described in this book are not related to individual components. Cost is an important factor of launcher competitiveness, and reducing component costs is an effective way to improve competitiveness. Therefore, we present a brief discussion about this topic at the end of this section.

Rocket's avionics tend to use high-grade components, which increase costs significantly. Consider relatively low-reliability components, which can be divided into three categories.

1. Factory defined aerospace grade. In principle, products are manufactured to the same standard as aerospace grade components, but the production line has not been verified by an appropriate aerospace organization. Product quality is high, but often with significantly reduced cost. Consumers can re-screen products if required.
2. Products are produced on the aerospace grade production lines, but only conventional screenings are performed. Consumers can perform supplementary screening. This grade is cheaper than factory defined aerospace grade, but it is important to note that component consistency is not the same as that for higher grade products.
3. COTS components. This includes industrial grade (−40 to +85 °C, plastic or ceramic packaging), wide temperature (−55 to +105 °C), and commercial grade (0 to +55 °C) components. These can also be employed through reasonable screening (called upgrading and screening) [54], derated design, redundancy design, etc.

The US Department of Defense, NASA, and ESA have developed specifications and standards for COTS applications [55, 56], which can be summarized as follows.

1. Define the environment (work temperature, humidity, salt fog, etc.).
2. Select components based on junction temperature rather than ambient temperature.
3. Degrade the components application conditions to meet the restrictions, including junction temperature requirements.
4. Design depending on real component characteristics rather than descriptions from the corresponding manuals.
5. Upgrade screening.

The main factors affecting reliability of COTS products include narrow working temperature range, plastic packaging, moisture sensitivity, and poor traceability of design, material, chip, and process control, etc. Therefore, upgrade screening should focus on these factors. However, screening is only effective for the specific task and lot, and is irrelevant if these components are intended for other tasks, or other grade components or lots are suitable for this task. Screening conditions are determined depending on component characteristics and applications.

Testing technology has changed greatly in recent years with ongoing design method and tool improvements. Application of integrated modular avionics (IMA) systems is increasing in aviation and aerospace industries. IMA provides a generic design platform, modular architecture, and onboard device can be integrated by module combinations. Based on the temporal and spatial partitioning operating system, application software is isolated from the hardware, and different applications are also independent. Following this trend, tests shall also take place modularized and hierarchically. If the design platform and modules are mature, testing could just focus on application and integration, not modules or boards, which can also reduce overall test costs. Interested readers are referred to [57, 58].

References

1. Xu YW (ed) (1992) Control system (2,3). In: Missile and aerospace collection - liquid ballistic missile and launch vehicle series. China Aerospace Publishing House, Beijing (in Chinese)
2. Hu HF et al (2010) Research on the test system and technology development of the launch vehicle control system. In: 3rd high level Forum on development strategy of experiment and test technology for national defense science and technology industry, Beijing, China, 12–13 Oct 2010 (in Chinese)
3. Cai YW, Yu GH (2004) Research on test method of launch vehicle. Aerosp Control 22(1): 55–57 (in Chinese)
4. Guo Y, He W, Xi H et al (2014) Research on optimization control technology in the launch vehicle ex-works testing state. Aeronaut Manufact Technol 16:99–103 (in Chinese)
5. Jin XS (2002) The environmental testing of spacecraft and its developing trends. Spacecraft Environ Eng 19(2):1–10 (in Chinese)
6. GJB1027A-2005 (2005) Test requirements for launch, upper-stage, and space vehicles (National military standard, in Chinese)

7. Zhang XD (2006) The revision of GJB 1027A–2005 "test requirements for launch, upper-stage, and space vehicles". Spacecraft Environ Eng 23(4):245–248 (in Chinese)
8. Jin XS (1999) Combined environment testing in the design of spacecraft. Environ Technol 2:14–22 (in Chinese)
9. Hu CH, Ma QL, Zheng JF (2010) Missile testing and launch control technology. National Defense Industry Press, Beijing (in Chinese)
10. Pressman RS (2011) Software engineering: a practitioner's approach, 7th edn of the original book, 1st edn of the Chinese version. China Machine Press, Beijing (in Chinese)
11. Myers GJ, Sandler C, Badgett T (2011) The art of software testing, 3rd edn. Wiley, Hoboken, NJ, USA
12. Chillarege R Software testing best practices. IBM, IBM research RC21457
13. Song ZY (2015) Embedded software testing technology for Chinese manned spaceflight project launch vehicle. In: Symposium on software engineering of Chinese space station project, Beijing, China, 5–6 Feb, 2015
14. Nidhra S, Dondeti J (2012) Black box and white box testing technique-a literature review. Int J Adv Comput Sci Appl 2(2):29–50
15. Kelly JH, Dan SV, John JC et al. A practical tutorial on modified condition/decision coverage. NASA, NASA/TM-2001-210876
16. Graaf B, Lormans M, Toetenel H (2003) Embedded software engineering: the state of the practice. IEEE Softw (11/12):61–69
17. Pontes RP, Martins E, Ambrósi AM et al (2010) Embedded critical software testing for aerospace applications based on PUS. In: XXVIII Simpósio Brasileiro de Redes de Computadores e Sistemas Distribuídos, XI Workshop de Testes e Tolerância a Falhas (WTF), Gramado, RS, Brazil, 24 a 28 de maio de 2010
18. Saxena A, Roychoudhury I, Lin W et al (2013) Towards requirements in systems engineering for aerospace IVHM design. In: AIAA infotech@aerospace conference 2013, Boston, Massachusetts, USA, 19–22 Aug 2013
19. MIL-STD-2165A: Testability program for systems and equipments,1993
20. GJB2547-95: Testability program for materiel, 1995 (National military standard, in Chinese)
21. QJ3051-1998: Testability design criteria for space products, 1998 (in Chinese)
22. Wang HJ (2008) Reviews of testability design technology. Sciencepaper Online 3(1):52–58 (in Chinese)
23. Li B, Zhang Q, Ren K et al (2010) Research on test design for spacecraft. Aerosp Control Appl 36(5):13–17
24. Nair C (2002) Modular test architectures for the aerospace industry. In: AUTOTESTCON proceedings, Huntsville, AL, USA, 17 Oct 2002
25. Fu P (2010) Development and suggestion of advanced test bus. In: 3rd high level forum on development strategy of experiment and test technology for national defense science and technology industry, Beijing, China, 12–13 Oct 2010 (in Chinese)
26. Han SW, Hu YH (1996) The application of VXI bus technology to aerospace ground test system. Missiles Space Veh 1:63–70 (in Chinese)
27. PXI™-5: PXI express hardware specification revision 1.0, PXI Systems Alliance, 2005
28. ANSI/VITA 46.0-2007: American national standard for VPX baseline standard. American National Standards Institute, Inc, 2007
29. Acromag. Introduction to VPX-VITA 46, 48, and 65: The Next Generation VME system replacement [DB/OL]. http://www.acromag.com/sites/default/files/Introduction_to_VPX_906A.pdf
30. ANSI/VITA 65-2010 (R2012) OpenVPX™ system specification. American National Standards Institute, Inc., 2012
31. Ken Grob. Using OpenVPX profiles to design, build systems [DB/OL]. Elma Electronic Inc. http://www.eetindia.co.in/STATIC/PDF/201203/EEIOL_2012MAR12_EMS_TA_01.pdf?SOURCES=DOWNLOAD
32. Poledna S (2010) TTEthernet communication. In: 58th IFIP working group 10.4 meeting on dependable computing and fault tolerance, Chicago, Illinois, USA, 24–27 June 2010

33. Steiner W, Bauer G (2008) Ethernet for space applications: TTEthernet. In: International spacewire conference, Nara, Japan, 4–6 Nov 2008
34. Steiner W, Maier R, Jameux D et al. (2008) Time-triggered services for spacewire. In: SpaceWire conference, Nara, Japan, 4–6 Nov 2008
35. TTEthernet—A powerful network solution for all purposes[DB/OL].TTTech Computertechnik AG. http://www.ttagroup.org/ttethernet/doc/TTEthernet_Article.pdf
36. Steiner W (2009) TTEthernet: time-triggered services for Ethernet networks. In: 28th IEEE/ AIAA digital avionics system conference, Orlando, FL, USA, 23–29 Oct 2009
37. Yi J, Xiong HG, He F et al (2014) Research on traffic classes transformation strategy and real-time guarantee scheduling algorithm in TTEthernet. Acta Aeronautica et Astronautica Sinica 35(4):1071–1078 (in Chinese)
38. Maguire LP, McGinnity TM, McDaid LJ (1999) Issues in the development of an integrated environment for embedded system design—part B: design and implementation. Microprocess Microsyst 23(4):199–206
39. Song ZY (2013) System synthesis technology based on HW/SW co-design for aerospace control system. Aerosp Control 31(2):9–15 (in Chinese)
40. Li LS (2012) Research on virtual vibration testing system of spacecraft. Harbin: Harbin Institute of Technology, July, 2012 (in Chinese)
41. Ricci S, Peeters B, Debille J et al (2010) Virtual shaker testing: a novel approach for improving vibration test performance. In: International conference on noise and vibration engineering, Leuven, Belgium, 20–22 Sept 2010
42. Ricci S, Peeters B (2009) Virtual shaker testing for predicting and improving vibration test performance. In: Proceedings of the IMAC-XXVII, Orlando, USA, 9–12 Feb 2009
43. Betts J, Vansant K, Paulson C et al (2008) Smart testing using virtual vibration testing. In: Proceedings of the 24th aerospace testing seminar (ATS), Manhattan Beach, USA, 8–10 Apr 2008
44. Appolloni M, Cozzani A (2005) Use of advanced integrated CAE tools to provide an end-to-end simulation of payload testing on Hydra. In: Proceedings of seminar on integrated open CAD/CAE platforms for mechanical engineering, ESTEC, Noordwijk, Netherlands, 28 Sept 2005
45. Kawatsu K, Taguchi H, Kurosu A et al (2015) Affordable high-reliability realization approach in liquid rocket engine development for new national flagship launch vehicle. In: TRISMAC 2015, Frascati, Italy, 18–20 May 2015
46. Naoki T, Nobuhiro Y, Akihide K et al (2011) An end-to-end high fidelity numerical simulation of the LE-X engine—engine performance evaluation and risk mitigation study. In: 47th AIAA/ASME/SAE/ ASEE joint propulsion conference & exhibit, San Diego, USA, July 31–Aug 3, 2011
47. Wang WQ, Liu C (2009) The simulation system for spacecraft acoustic test. Spacecraft Environ Eng 26(2):140–142 (in Chinese)
48. Dou Q, Pang HW, Wei CF et al (2007) The software architecture of spacecraft virtual thermal testing platform and its application. Spacecraft Environ Eng 24(6):370–373 (in Chinese)
49. Brinkmann PW, Kretz D (1992) The design concept of the 6-degree-of-freedom hydraulic shaker at ESTEC. In: The 17th space simulation conference on terrestrial test for space success, Goddard Space Flight Center, USA, 1 Nov 1992
50. Qian ZY, Xiao W, Xu LJ et al (2012) Application of force limited method in vibration test of spacecraft. Spacecraft Eng 21(3):101–106 (in Chinese)
51. Hobbs GK (2012) HALT and HASS, accelerated reliability engineering (trans: Ding QB). HALT and HASS accelerated reliability engineering. Beijing, Aviation Industrial Press, June 2012 (in Chinese)
52. Chu WH, Chen X, Tao JY et al (2002) Highly accelerated life test (HALT) and highly accelerated stress screen (HASS). Struct Environ Eng 29(4):23–37 (in Chinese)
53. Li M, Chen YX, Kang R (2011) Performance modeling and consistency test for inertial components in high accelerated stress screening. J Chin Inertial Technol 19(1):111–115,126 (in Chinese)

54. Electrical, electronic, and electromechanical (EEE) parts upgrade screening and qualification requirements revision E. Johnson Space Center, Houston, USA, SSQ25001
55. Teverovsky A, Sahu K PEM-INST-001: instructions for plastic encapsulated microcircuit (PEM) selection, screening, and qualification. NASA, NASA/TP-2003-212244
56. General requirement for the use of commercial EEE parts in space applications. French Space Agency, RNC- CNES-Q-60-523; German Space Agency, DLR-RF-PS-006
57. Prisaznuk PJ (1992) Integrated modular avionics. In: Aerospace and Electronics Conference, Dayton, OH, USA. 18–22 May 1992
58. Tagawa GBS (2011) Marcelo Lopes de Oliveira e Souza. An overview of the integrated modular avionics (IMA) concept. In: Brazilian conference on dynamics, control, and applications (DINCON 2011), Águas de LIndóia, Sao Paulo State, Brazil, 29/08 a 02/09/2011

Chapter 3
Equipment Level Test

This section introduces equipment level testing, which incorporates functional/performance and reliability testing.

An important guideline in aerospace is *test what you fly*, and *fly what you test* [1]. The first phrase emphasizes test equivalence, i.e., reduces any difference between test conditions on the ground and in-flight, the functions used in-flight shall be tested on the ground. However, many test cases are difficult to perform during system level testing, and equipment level testing has become an important approach to ensure test effectiveness.

Debugging is often regarded as a test, where if each required function has been successfully debugged, the test is considered finished. However, debugging is performed under static stimulus to check static response, and the environment and conditions are inconsistent with in-flight conditions. Thus, the test strength is insufficient, and boundary conditions in particular are not checked. Therefore, equipment level testing must cover real working or stricter conditions wherever possible. Consider the features and complexity of control system equipment, we introduce two equipment categories: controllers, including the OBC; and inertial devices, including the IMU and other devices composed of accelerometers and rate gyros. Other devices, such as power distribution devices, bus networks, are not discussed here.

This section also considers reliability testing, including reliability growth and enhancement tests. We discuss revising reliability growth test conditions depending on practical in-flight measurements, and calculating reliability testing parameters. The reliability enhancement test's purpose is to quickly identify product defects, providing the basis for improvements, and to obtain the condition limits the product can bear, which form the basis for HASS.

Z. Song, *The Test and Launch Control Technology for Launch Vehicles*,
https://doi.org/10.1007/978-981-10-8712-7_3

3.1 Functional and Performance Tests

Functional and performance testing is the basis of equipment level testing and forms the fundamental basis for all testing where the UUT is powered. Functional and performance testing at the equipment level, sometimes referred to as unit testing, should focus on the following aspects when designing test cases.

1. Simulate actual conditions as much as possible.
   For example, OBC samples gyro and accelerometer information through multichannel pulse counter interfaces. If each channel is tested individually, the test case fails to cover simultaneous sampling on all channels, and in-flight circuit EMC characteristics are insufficiently examined.
2. Ensure that test cases cover all task requirements.
   Set up a requirement tracking table, as shown in Table 3.1, where the first column contains the requirement descriptions. To avoid missing something, each requirement entry should be listed: if an entry is descriptive, i.e., does not correspond to specific test requirements, an explanation can be added in the second column. This table will be fed back to those drafting requirement descriptions, and confirmed by both sides, particularly to check if entries with explanations actually cannot be checked by any test requirement. This also helps to remind the requirement presenter to improve descriptions so only testable entries are described.
3. Consider both functional and performance testing.
   Performance tests are usually performed during debugging, whereas only functional tests are performed out automatically in equipment level acceptance test. This traditional approach retains hidden problems.
   Many performances are measurable, but are just not as convenient to be checked as functional testing. Consider an avionics device for example. Features of real-time, time precision, maximum load rate, maximum communication rate, anti-interference threshold, interrupt response time, etc. can be examined for at least some cases. Thus, the quantitative indexes would not be evaluated if there were no performance testing.
4. All possible working conditions should be examined.
   Normal and boundary conditions should be considered, where normal refers to operation within nominal conditions. However, all quantitative indexes have boundary conditions, such as upper and lower power supply voltage limits, maximum communication rate, maximum communication distance, number of concurrent interrupt limits, threshold voltage to determine effective signals, etc.

Table 3.1 Test requirement tracking table

Requirement description	Test requirements	Test case	Expected test results	Other supporting measures
…	…	…	…	…

Thus, boundary conditions for these indexes should be simulated even stricter than actual constraints to check if the design has sufficient margins.
Tests should be performed under normal and fault conditions. Fault conditions mean some faults are simulated to check if the UUT can adapt. There are two types of faults: input signal or load devices, and the UUT itself. The simulated faults should not damage or adversely affect the UUT. Fault simulation is essential for redundant designs to verify correct fault detection and isolation (FDI) function.
Tests should be performed under static and dynamic (transient) conditions. Static testing is common, but performance checks under dynamic conditions are more critical, and operational conditions in-flight are strongly dynamic. For example, the dynamic index of an inertial device is completely different from the calibration result under static conditions. A transient state refers to instantaneous or transition conditions of electrical signals, such as the amplitude and duration of peak surge current, which will influence power source capacities and EMC performance.

5. Integration test of software and hardware.
   The final version software and hardware of products with embedded software should be integrated to check compatibility, which cannot be replaced by other debugging methods. It is also generally unacceptable to substitute software/hardware integration with system level testing, because the latter is often significantly costlier.

3.1.1 Controller Testing

The controller includes the OBC, integrated (ICU), and servo (SCU) controllers. Their basic configurations are similar with platform design technology, which is convenient for testing.

Each controller comprises a processor, power supply, and I/O modules. The processor module is generally a small computer system, including a CPU, memory, logic control circuit, and internal bus interfaces. The I/O modules vary depending on the specific function; for example, OBC I/O modules generally comprise bus interfaces (1553B, 422, 485, LVDS, etc.) and discrete control interfaces, whereas ICU I/O modules include high-power discrete output interfaces, SCU I/O modules include analog amplifying circuits (to drive servo-valves), and thrust regulation controllers I/O modules include pulse-width modulation (PWM) control interfaces.

Consider the ICU test scheme as a typical example. When the ICU receives a shutdown command from the OBC through a 1553B bus, or samples exhaust signals, it sends sequential cutoff signals automatically. Each discrete control signal drives electromagnetic valves and pyrotechnics, and the signals are grouped based on current load requirements. For example, the valve driving current should not exceed 7 A, whereas pyrotechnic ignition current should not exceed 50 A for

50 ms. The ICU also samples triple modular redundancy (TMR) pressure signals through the RS-485 interface, filters these depending on the pressure control algorithm, and controls the corresponding electromagnetic valve. Besides, it collects the responses of these control signals for BIT and transmits BIT data to the bus monitor (telemetry system). After power on, the ICU completes a self-check and repeats this whenever it receives the *self-inspection* command from the OBC, returning the check results to the OBC. The debug interface is retained to upload software or re-code programmable components without opening the chassis. The ICU is powered by 28 V and TMR configured.

Test requirements are dictated by the ICU functions, and the test facility is shown in Fig. 3.1 for a simulated operation environment, as follows:

1. Discrete output for reset, takeoff, shutdown, exhaustion signals, etc.
2. Two 1553B terminals representing the OBC and bus monitor (MT).
3. Multiple RS-485 interfaces to simulate pressure sensors.
4. Test interfaces for sequential control signals, such as engine ignition, cutoff, pyrotechnics detonation, pressure control, etc.

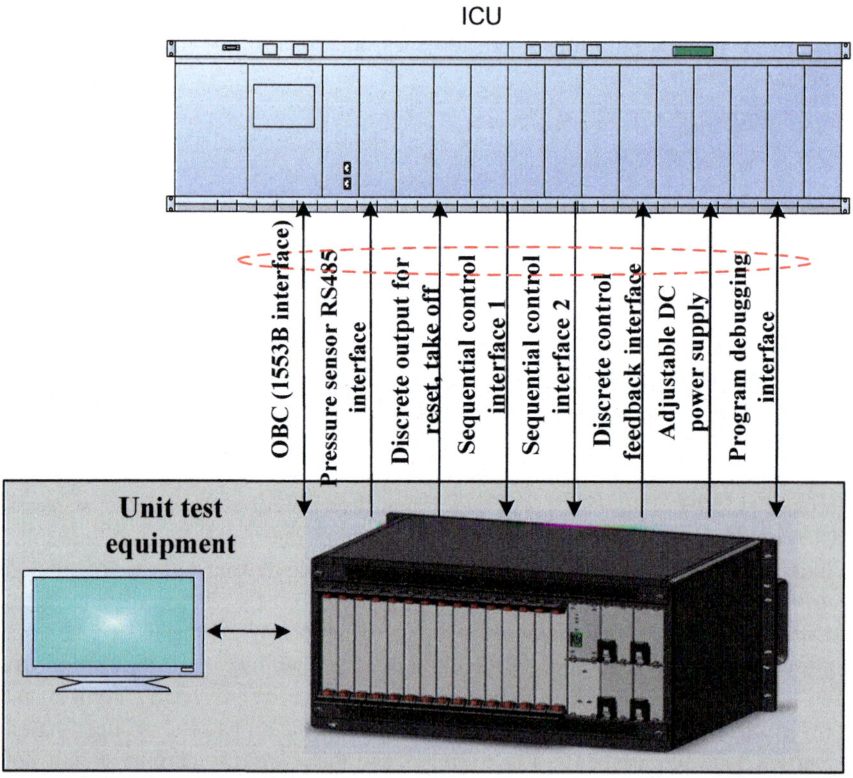

Fig. 3.1 ICU test facility diagram

5. Load simulations for valves and pyrotechnics, so discrete control signal feedback can be collected.
6. Debug interface.
7. Adjustable power supply interface, to regulate voltage to the upper or lower working limits.

Depending on the working scenario, the test facility simulates the takeoff signal to start the test, all other required commands, and signals; and collects all the feedback or responses, to verify ICU operation. Reliability test requirements, signal characteristics, fault simulation, and boundary conditions are considered in addition to the functional tests described above.

1. Each module in TMR configurations should be separately powered on or loaded with software, which provides a convenient way to simulate a given module failure, such as power off or software unloading, to check redundant management.
2. Abnormal information messages should be simulated to check fault detection.
3. Triple redundant pressure sensors and data bus outputs can be set differently and asynchronously to check ICU adaptability under abnormal conditions.
4. Abnormal pressure scope and trend should be included in the pressure simulation. Other simulation cases include starting or stopping pressurization, retaining the current state, communication breakdown, etc. Setting a fixed value is strongly not recommended.
5. Power voltage should be regulated, e.g., for 25–31 V.
6. The pressure sensor communication cycle should be regulated consistent with sensor timer accuracy, e.g., 18–22 ms.
7. Discrete signal amplitudes, such as takeoff and exhaust, should be adjustable to check command thresholds are set and processed correctly.
8. Maximum static current for each discrete control channel should be detectable, e.g., 7 A for valve control.
9. Maximum pulse current for each discrete control channel should be detectable, e.g., 50 A for 50 ms for pyrotechnics control.
10. Transient power-on conditions, including surge current magnitude and duration, should be monitored

Items (9) and (10) can be performed in routine testing, but it is important that the maximum static current be included in burn-in and reliability growth testing, with load connected. Item (9) checks overload output capacity, and there are usually restrictions on the number of action times.

Test facility software must ensure real-time operation, so one of the following architectures is adopted.

1. Host (upper) and slave (lower) computers, with an embedded real-time OS (RTOS) on the slave computer to ensure real-time requirements, and a commercial OS, such as Windows, on the host computer for user-friendly interfaces. Host and slave computers should be networked.

2. A single computer employing a real-time extension (RTX) embedded Windows OS [2], with the user interface running in the traditional Windows environment, and real-time applications executing in the RTX environment.

Clearly, equipment testing is not simple. If the UUT varies, test facilities must also vary correspondingly and be customized, with significant costs. Thus, platform design is preferable, which facilitates unit testing substituted by module testing in advance, and integration testing is greatly simplified by design platform maturity and module reuse. Similarly, the test facility can be in common use.

3.1.2 *Inertial Device Testing*

Gyro and accelerometer outputs for different positions and during turning processes can be obtained from high-precision positioning and rate tables, so equations for meters error coefficients are solved. These coefficients are stored in the OBC for in-flight navigation. However, inertial device accuracies under dynamic or vibration conditions are often significantly different from the static calibrated results [3]. Although the coefficients cannot be separated under dynamic conditions, some new methods have been developed to verify dynamic accuracy.

Inertial device calibration has been fully discussed elsewhere [4, 5], so only dynamic accuracy verification methods are proposed here. The method examines IMU accuracy under the application conditions, which is comprehensively different from calibration conditions, and functional testing, where the performance or accuracy cannot be assessed. Considering the IMU experiences many vibration conditions in-flight, accuracy evaluation in these scenarios have been paid increasing attention. This section introduces two verification methods: navigation calculation during vibration and comparison of redundant units.

3.1.2.1 Navigation Accuracy Under 6-DOF Vibration

In-flight vibration includes linear and angular vibration, and Fig. 3.2 shows a typical test system simulating these conditions.

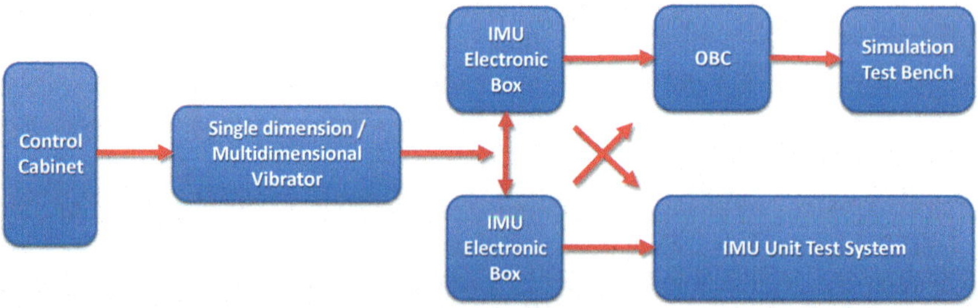

Fig. 3.2 Navigation accuracy test system block diagram

The IMU is installed on the vibration table, and its outputs under 6-DOF vibration conditions are collected and navigational calculations performed. Since the IMU is in static state and the same location before and after vibration, and attitude after vibration is restored to the original state, so the deviations of IMU's velocity, position, and attitude before and after vibration should be zero. This is the theoretical value for the calculated result from the navigation algorithm. Considering random meter errors, if the deviations exceed a given threshold, the calibration parameters cannot hold for IMU performance under the applied vibration. Therefore, analysis can check or locate which meter(s) significantly affect navigational violation(s). For example, if the deviation of pitch attitude angle is very large, then the Z-axis gyro should be inspected, whereas if the roll attitude angle changes significantly, the X gyro requires evaluation. Pitch attitude and Y accelerometer affect X-axis position deviation, while yaw attitude and Z accelerometer mainly affect Z-axis position deviation, etc.

However, it is difficult to accurately distinguish sensible in-flight ratios for angular and linear vibrations, so proportions are assigned to the vibration conditions based on engineering experience, as shown in Table 3.2 for example.

Based on Table 3.2, proportions were assigned as 1:9 for angle and linear vibration, respectively, with −9, −6, and −3, and 0 dB magnitude, and application time = 1 min. Sometimes, the 2:8 ratio is adopted.

The process is shown in Fig. 3.3. In this example, navigation time was 600 s. Considering this IMU accuracy, navigation error thresholds for positions along X-, Y-, and Z-axes were 1000, 800, and 1000 m, respectively.

The method can effectively expose the IMU weaknesses, allowing continuous accuracy improvement. The influence on accuracy during vibration is significant:

Table 3.2 Linear and angle vibration experiment condition

Installation position	Frequency range (Hz)	Qualification level		Acceptance level	
		PSD (g^2/Hz)	Total RMS acceleration (g)	PSD (g^2/Hz)	Total RMS acceleration (g)
Payload bracket, onboard device mounting in instrument cabin	20–150	6 dB/oct	16.92	6 dB/oct	8.46
	150–800	0.2		0.05	
	800–2000	−3 dB/oct		−3 dB/oct	

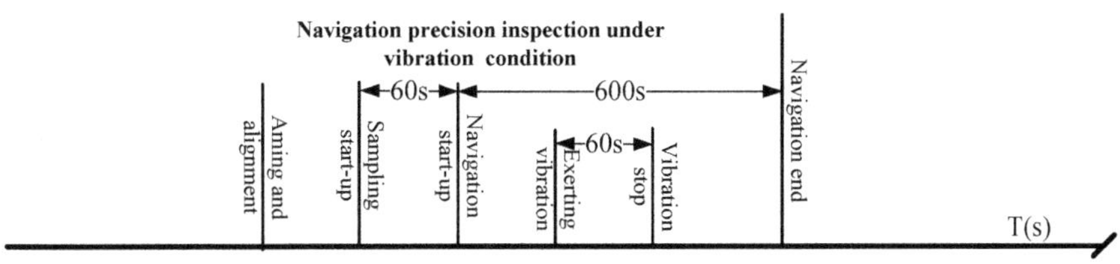

Fig. 3.3 6-DOF navigation accuracy vibration experiment process

when the vibration extends for 10 min or reaches magnitude 0 dB, both navigation results exceed their thresholds. However, there is no clear pattern for the influence of linear and angular vibration ratios for different IMU types or various products of the same type.

Although navigation deviations may exceed the thresholds, large vibrations mainly occurred for engine ignition, shutdown, and stage separation, which have shorter duration than test vibration time. Thus, deviations under vibration do not generally produce in-flight accuracy violations. The main purpose of the vibration test is to identify IMU weaknesses, and under appropriate test conditions, it may become an identification activity or acceptance test. Furthermore, the test outcomes remind us that we cannot simply use calibration parameters to evaluate in-flight IMU performance.

3.1.2.2 Redundant Onboard IMU Consistency

It is difficult to evaluate IMU accuracy once it is installed. However, if triple redundant IMUs are installed, FDI technology used in-flight can be adopted for testing. Generally, all IMUs are installed in the same holder, eliminating installation error influences, such as levelness and launch direction. Hence, differences between IMU sensing can be used to verify the accuracy of the other.

Consistency thresholds can be far stricter during ground testing than in flight, because there is no overload and vibration. The IMU can only sense local gravity and Earth's rotation, so error thresholds on ground test can be accurately determined.

Range analysis can be used for evaluation. Consider the gyro for example. Errors between two measurements along the same axis should not exceed 6σ, where σ = the gyro standard deviation, and 3σ is the acceptable maximum deviation for one meter. For extreme cases, one gyro's error might = $+3\sigma$, where another's = -3σ, producing a difference of 6σ. If difference $>6\sigma$, for whatever reason, this is considered a violation. However, if error $<6\sigma$, this does not necessarily mean gyros are acceptable. Thus, range analysis actually relaxes the examination requirement. Although it is more reasonable to determine thresholds based on the probability distribution of the difference of two random measurements, the range analysis is significantly easier and simpler. When a violation is identified, the failure isolation process is the same as in-flight [6].

3.2 Reliability Growth and Enhancement Testing

Most failures are related to environmental factors, with vibration, impact, noise, etc. during ascent phase having a significant influence on space product reliability. Many space agencies have summarized environmental test methods to expose failure modes and mechanisms under different environmental stresses, to provide the basis for improvement [7, 8]. Current environmental reliability testing is focused on integrating stress simulation developed from single to multiple stresses, such as temperature, humidity, vibration, electrical, etc., commonly called reliability growth testing. The environmental test evolves from low to high accelerated stress levels, i.e., reliability enhancement testing; from single device to system level testing, as discussed in Sect. 2.4. The following sections discuss reliability growth and enhancement testing.

3.2.1 Test Condition Determination

Most LV electronic device is installed in the instrument cabin and interstages between fuel tanks. We select the representative installation position in the center of the instrument cabin as an example to compare vibration conditions for reliability testing and that of flight [9], as shown in Fig. 3.4.

Vibration test condition for low-frequency sections covers in-flight outcomes; vibration test conditions differ significantly from the in-flight environment beyond 200 Hz. Therefore, reliability growth test conditions should be corrected depending on in-flight sensor information, using the following principles.

1. If in-flight conditions do not exceed test conditions, test conditions should not change.
2. If in-flight conditions exceed test conditions, then test conditions of these frequencies should cover the in-flight conditions.

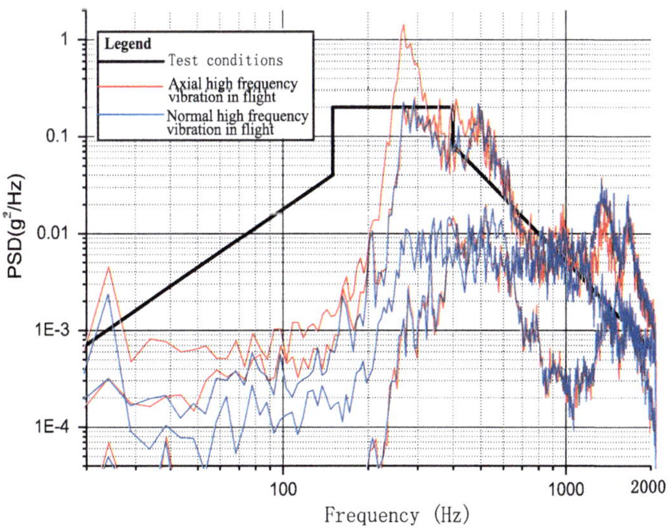

Fig. 3.4 Test conditions and in-flight high-frequency vibration for instrument cabin disk platform

3. Test conditions should just fit in-flight conditions, with relatively small margins to control the total root-mean-square (RMS) acceleration during test.

Random vibration stress in the reliability growth test is controlled by the acceleration power spectrum density set in advance, which is usually expressed in decibels corresponding to the different frequency ranges. The acceleration power spectrum density (PSD), $w(f)$, represents the mean square value of the random signal $x(t)$ passing through the narrow band filter with central frequency f and a specific bandwidth. Figure 3.5 shows an example of the PSD calculation for a typical random vibration condition.

The ascent spectrum can be expressed as

$$w_2 = w_1 \cdot \left(\frac{f_2}{f_1}\right)^{\frac{A}{10\lg 2}} = w_1 \cdot \left(\frac{f_2}{f_1}\right)^{m_a}, \tag{3.1}$$

where f_1 and f_2 are the start and end frequencies of the ascent spectrum, respectively; w_1 and w_2 are the PSDs for these frequencies, respectively, as shown in Fig. 3.5; A is the ascent slope (dB/oct); and

$$m_a = \frac{A}{10\lg 2} \approx \frac{A}{3}.$$

Similarly, the descent spectrum can be expressed as

$$w_4 = w_3 \cdot \left(\frac{f_3}{f_4}\right)^{\frac{B}{10\lg 2}} = w_3 \cdot \left(\frac{f_3}{f_4}\right)^{m_d}, \tag{3.2}$$

where f_3, f_4, w_3, and w_4 are the start and end frequencies and the corresponding PSDs, respectively, of the descent spectrum, as shown in Fig. 3.5; B is the ascent slope (dB/oct), and

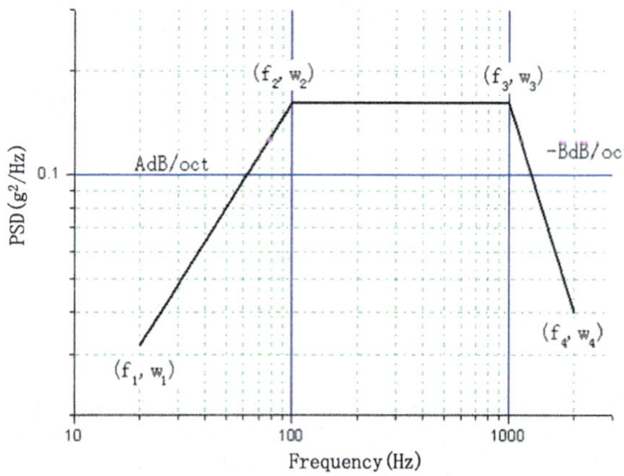

Fig. 3.5 Example of random vibration power spectrum density

$$m_d = \frac{B}{10\lg 2} \approx \frac{B}{3}.$$

Total mean square acceleration in the reliability growth test represents the total vibration magnitude, i.e., the total energy, applied to the UUT. And G_{rms}, the total RMS acceleration for a random vibration between frequency point f_s and f_e, is calculated:

$$G_{\text{rms}} = \left(\int_{f_s}^{f_e} w(f)\mathrm{d}f \right)^{1/2}. \tag{3.3}$$

Thus, RMS acceleration is determined by the area enclosed by the PSD between the frequency ranges. The calculation usually includes three parts: horizontal, ascending, and descending spectra, as shown in Fig. 3.6.

The horizontal spectrum can be expressed as

$$G^2_{\text{rms},h} = w_{i,h}(f_{j,h} - f_{i,h}), \tag{3.4}$$

where f *and* w represent the frequency and corresponding PSD, suffixes i and j represent the start and end frequency identifiers in the spectrum, and suffix h represents the horizontal spectrum.

The ascending spectrum is

$$G^2_{\text{rms},a} = \frac{w_{j,a} \cdot f_{j,a}}{\frac{m_a}{3} + 1} \left[1 - \left(\frac{f_{i,a}}{f_{j,a}} \right)^{\frac{m_a}{3}+1} \right], \tag{3.5}$$

where suffix a represents the ascending spectrum; and the descending spectrum is

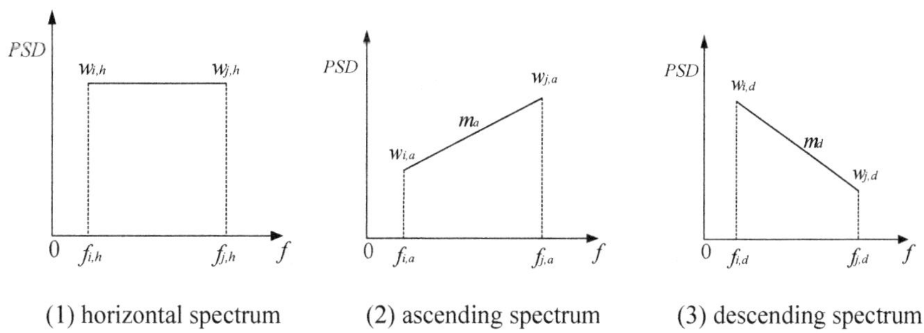

Fig. 3.6 Typical spectrum

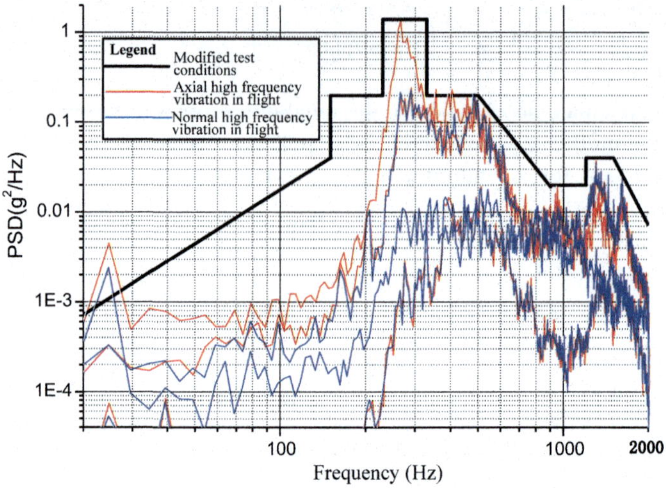

Fig. 3.7 Correction of the test condition for instrument cabin disk platform

$$G_{\mathrm{rms},d}^2 = \frac{w_{i,d} \cdot f_{i,d}}{\frac{m_d}{3} - 1}\left[1 - \left(\frac{f_{i,d}}{f_{j,d}}\right)^{\frac{m_d}{3}-1}\right], \tag{3.6}$$

where suffix *d* represents the descending spectrum.

Thus, the total RMS acceleration is the sum of the segments,

$$G_{\mathrm{rms}}^2 = G_{\mathrm{rms},h}^2 + G_{\mathrm{rms},a}^2 + G_{\mathrm{rms},d}^2 + \cdots \tag{3.7}$$

Figure 3.7 shows correction to in-flight conditions for this example.

3.2.2 Reliability Growth Testing

The reliability growth test verifies product reliability, identifying any weakness through fault analysis. The test is repeated after improvements to obtain the reliability growth [10]. In-flight reliability will require many environmental test cycles, with UUT function and performance evaluated in each cycle. Environmental factors include temperature, humidity, electrical, and vibration stress. Figure 3.8 shows the control block diagram for the reliability growth test system.

Temperature is automatically controlled during testing in the environmental testing chamber. Driving signals are sent by the vibration controller to produce random vibration based on the specific power spectrum, and amplified to drive the vibration table. Sensors on the table feed the response back, creating closed-loop vibration control to ensure table vibration responses, are consistent with test

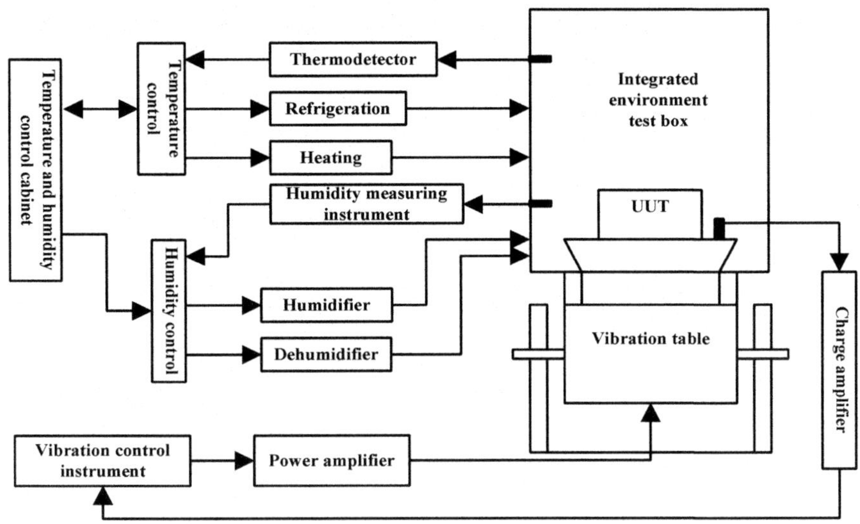

Fig. 3.8 Control block diagram for reliability growth test system

requirements. Sensors on the UUT provide the response spectrum of their installation point. The UUT is fixed on the table using an inverted T-type clamp within the test chamber, while the vibration table itself is beneath the chamber.

Figure 3.9 shows a typical test site. Vibration control is based on averaging three sensors' outputs (labeled 1#, 2#, and 3#), and sensor 4# is set where UUT stiffness is larger and near the damper, with its output considered as the UUT vibration response. To avoid vibration amplification, the vibration control spectrum is generally determined before the formal test. Figure 3.10 shows a typical test profile.

Power-on time, number of test cycles, and random vibration timing for the reliability growth test are calculated depending on the following schemes.

1. Total power-on time.

Total UUT power-on time, T_{pon}, is determined from the fixed time exponential truncated solution,

$$T_{\mathrm{pon}} = \frac{1}{n} \cdot \left[-\frac{\chi^2_{\gamma,2f+2}}{2\ln R_{\mathrm{L}}} t_{t0} \right], \tag{3.8}$$

where R_{L} is the reliability growth target value; γ is the confidence degree; f represents failure number and is set as 0; $\chi^2_{\gamma,2f+2}$ is the lower quantile of the χ^2 distribution with confidence degree γ, e.g., $\chi^2_{0.7,2} = 2.41$; t_{t0} is the temperature cycle task time, for onboard avionics, it is working time in-flight; and n is the UUT numbers.

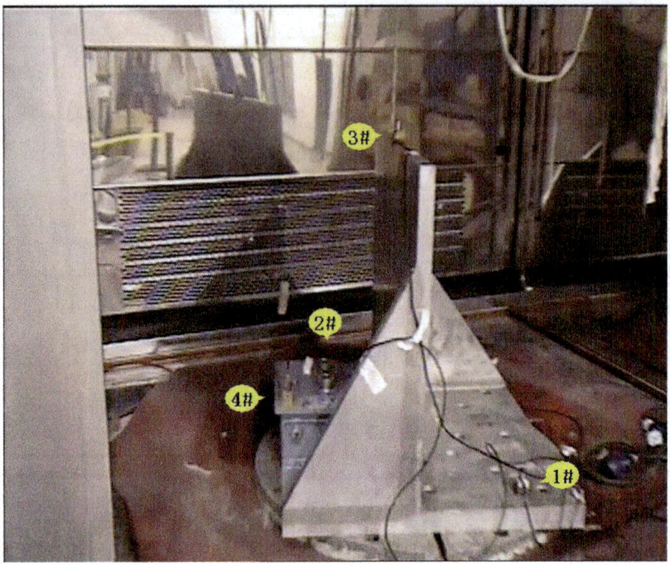

Fig. 3.9 Reliability growth test tooling and measuring points

2. Number of test cycles.

The number of test cycles for a given UUT, N, is calculated as

$$N = \frac{T_{\text{pon}}}{T_{0\,\text{pon}}}, \tag{3.9}$$

where $T_{0\ \text{pon}}$ is power-on time in a single temperature cycle.

3. Random vibration time.

Calculation of the random vibration test time is separated into two cases depending on the total RMS acceleration at the electronic devices installation site.

i. If $G_{\text{rms}} > 10g$, the product total random vibration time in each direction, T_{V}, is determined by the Weibull distribution,

$$T_{\text{V}} = \left(\frac{\ln(1-\gamma)}{n \ln R_L}\right)^{\frac{1}{m}} t_{v0}, \tag{3.10}$$

the vibration stress should be applied within the first cycle. If that is too difficult, the stress should be relatively concentrated and allocated equally in previous several cycles.

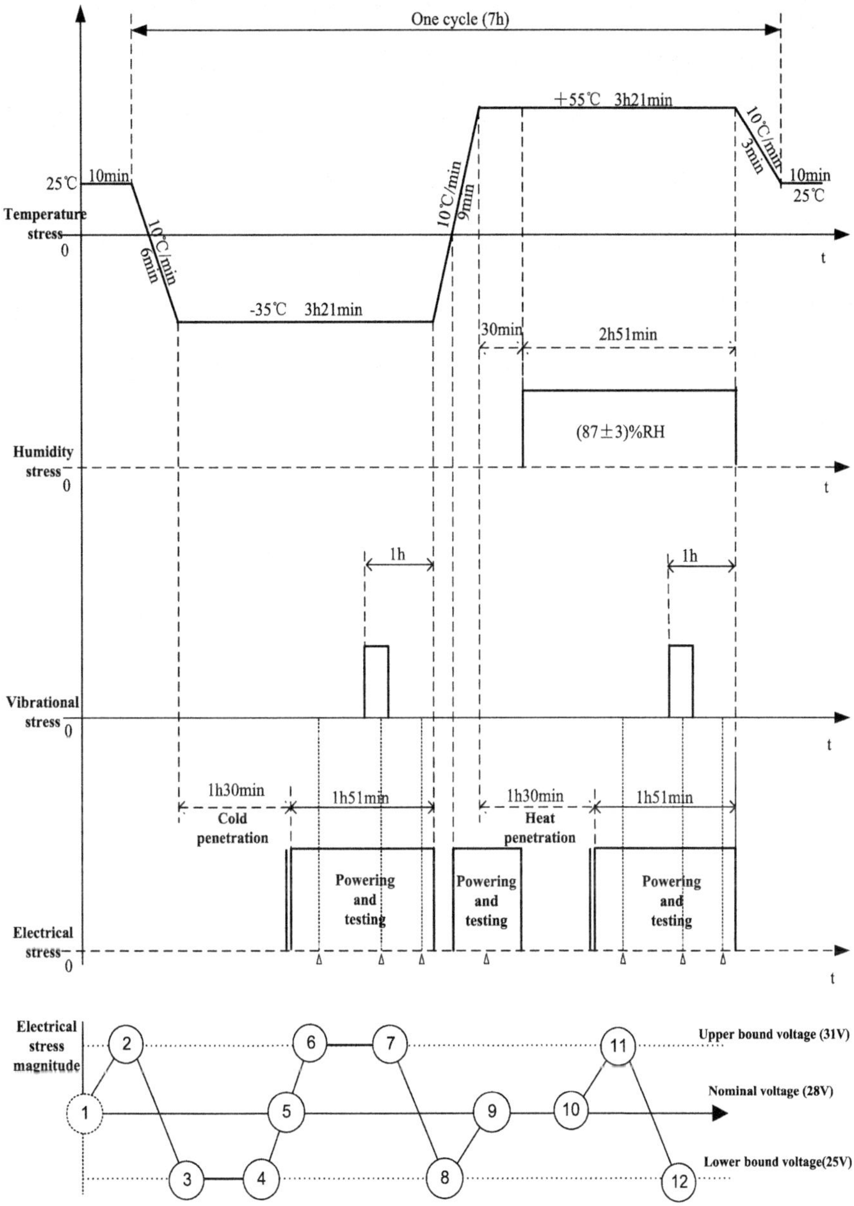

Fig. 3.10 Typical test profile

ii. If $G_{rms} \leq 10g$, T_V, is determined by the exponential distribution,

$$T_V = \frac{1}{n} \cdot \left[-\frac{\chi^2_{\gamma,2f+2}}{2 \ln R_L} t_{v0} \right], \tag{3.11}$$

the vibrational stress may be applied in the first cycle or equally assigned in previous cycles.

Here, m is Weibull distribution shape parameter, and set as 1.2 for control system electronic devices; t_{v0} is the random vibration task time, for onboard avionics, it is generally set as 1 min.

3.2.3 *Reliability Enhancement Testing*

Reliability enhancement testing rapidly activates design and technique defects of the product, exposing any weaknesses. Failure cause and mode analysis then allows appropriate improvements to shorten the development cycle. The UUT is placed in a harsh environment and rigorous functional state, and the stimulus magnitude gradually increased until failure occurs. These environmental and functional states may not correspond to service conditions; they are only used to rapidly stimulate weakness.

The following sections discuss step and comprehensive reliability enhancement test methods.

3.2.3.1 Step Tests

This method is also known as HALT, where the UUT stress limit is determined step-by-step from the normal service state. Stimuli are divided into temperature step, temperature cycle step, step vibration, and combined temperature and vibration step tests [11, 12]. Three-axial 6-DOF random vibration is preferred for HALT vibration testing, but can also be substituted with single axis random vibration tests. The tests usually abort when the desired steps are completed or the UUT is damaged, but can continue until the UUT breaks to find the maximum tolerable boundary stress, and can resume after the UUT is improved or repaired after breakdown. When stress reaches the UUT tolerance limit, the test can be terminated without requiring repair.

Temperature Step Stress Test

Select cooling and heating step, and gradually lower or raise the temperature until the UUT fails or the desired minimum (T_{LG}) or maximum (T_{HG}) is achieved. Figures 3.11 and 3.12 show typical test profiles using 5 °C increments.

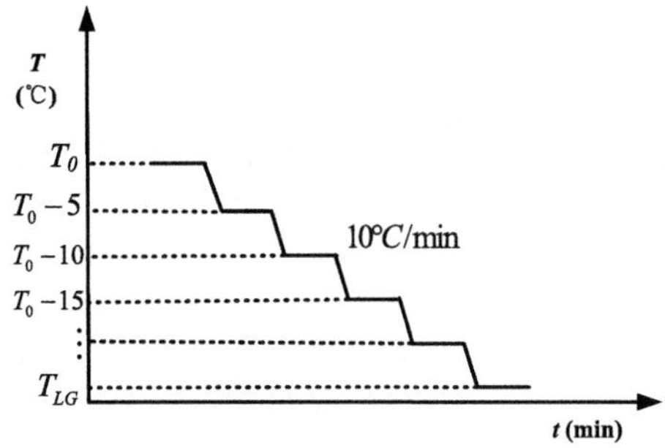

Fig. 3.11 Cooling step stress test profile

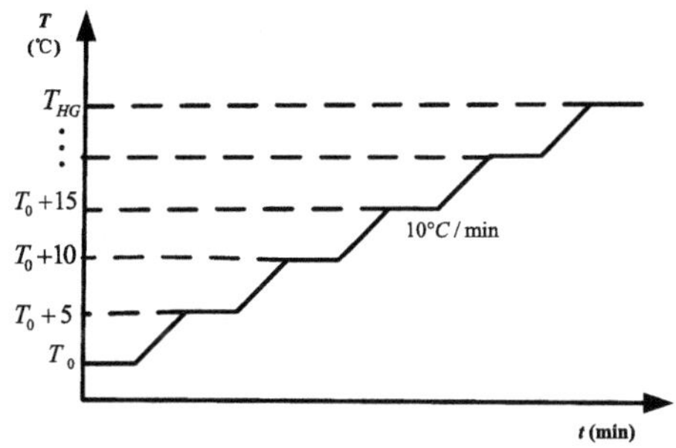

Fig. 3.12 Heating step stress test profile

The initial temperature, T_0, is set as the minimum or maximum temperature during screening, acceptance, and identification tests. Temperature change rate should be no less than 10 °C/min, and the holding time should meet the UUT cooling or heating requirements plus working time. It is important that the UUT works when thoroughly cooled or heated, with 50% of operating time under nominal electrical stress (voltage), with 25% of the time at the highest and lowest electrical stress. If not specified, the maximum or minimum voltage variation range is 10% above or below the rating.

Temperature Cycling Step Stress Test

The test stimulates product failure by increasing the rate of temperature change, with maximum = $T_{HG} - 5$ °C and minimum = $T_{LG} + 5$ °C. Initial temperature change rate, ΔTG, is 10 °C/min, increasing by 5 °C/min each cycle. The test continues until product failure or test conditions reach the pre-set maximums. Figure 3.13 shows a typical test profile.

Step Vibration Test

The initial vibration total RMS magnitude is set at the highest value found through previous environmental stress screening, vibration acceptance, and qualification

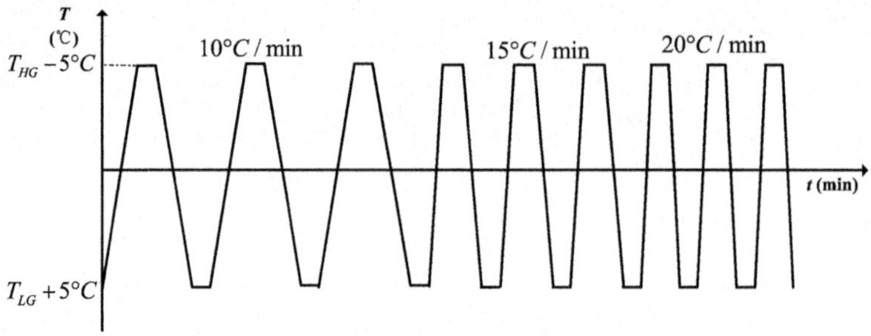

Fig. 3.13 Temperature cycling step stress test profile

Fig. 3.14 Step random vibration test profile

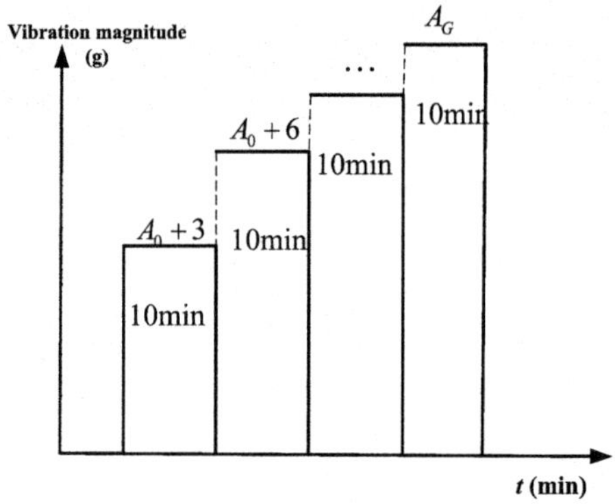

Fig. 3.15 Step sine scanning vibration test profile

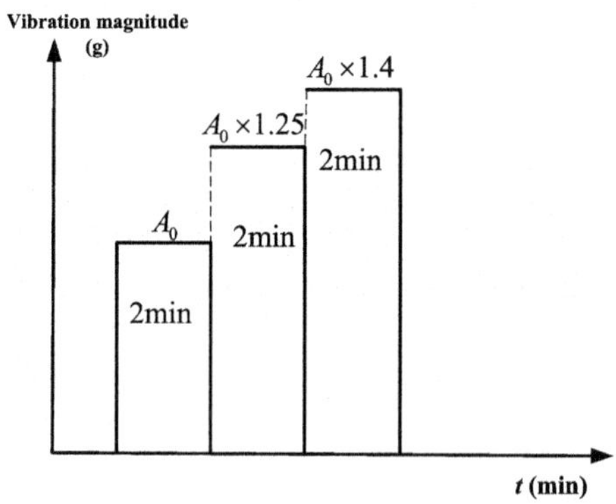

tests. Total RMS is then gradually raised in fixed steps for the random vibration test, whereas the magnitude increases by proportion for the sine scan vibration test. Figures 3.14 and 3.15 show typical step random vibration and sine scan vibration test profiles, where the UUT is powered on for 10 min.

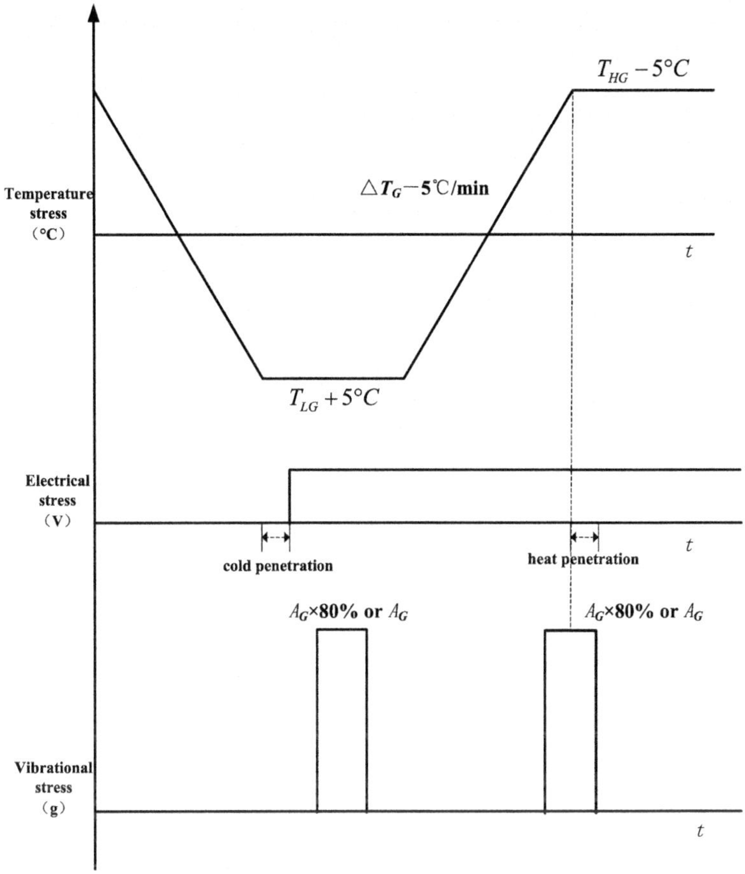

Fig. 3.16 Temperature and vibration combined environmental test profile

Combined Temperature and Vibration Test

This test applies temperature and vibration stress simultaneously, as shown in Fig. 3.16.

3.2.3.2 Comprehensive Reliability Enhancement Test

This is a 3-axis 6-DOF, broadband pseudorandom vibration and large temperature change rate cycle test, to rapidly stimulate potential defects, also known as HASS. The tolerable stress limit refers to HALT, and a slightly lower stress level is chosen as the HASS basis. HASS stress may exceed UUT service conditions, but since this is a product screening test, it should not cause damage or produce new defects, nor excessively consume the effective product life. HASS is the fastest and most effective screening process.

The 3-axis 6-DOF random vibration refers to linear acceleration vibration generated along and rotational vibration around three orthogonal axes (X, Y, and Z). The test facilities comprise a high acceleration test chamber, compressed air system, and liquid nitrogen auxiliary refrigeration system. The compressed air system operates the air hammer striking the test table to produce multi-axis continuous non-Gaussian broadband pseudorandom vibration. The peak probability density distribution is no less than 10σ, significantly larger than that generated by traditional electromagnetic vibration tables, which have only 3σ. Hence, the efficiency to stimulate defects is very high. The liquid nitrogen refrigeration system allows rapid cooling of the chamber.

This is a comprehensive environment test, with a wider frequency band and temperature range, and faster temperature change rate than general test systems. Typical test parameters are as follows:

1. Acceleration = 10–5000 Hz, 0–90 g (RMS).
2. Temperature = −100 to +200 °C.
3. Temperature change rate >60 °C/min.

Figures 3.17 and 3.18 show typical test facilities and corresponding PSD, respectively.

The maximum frequency is 1000 Hz in Fig. 3.18a, and the magnified range in Fig. 3.18b shows the frequency range of 1000–5000 Hz. The frequency range of most traditional motor-driven vibration tables is less than 2000 Hz (Fig. 3.18b, vertical line). Multi-axis vibration provides more energy at high frequencies, but reduced energy at low frequencies. The abundant high-frequency modes in this vibration type cause higher stresses within the unit test time, quickly producing fatigue. Therefore, high-frequency sensitive components, such as surface mount technology (SMT) internal and external welding, are susceptible to this test, whereas low-frequency vibration is more effective for large components. This test is performed without dampers, since they strongly restrain high-frequency vibrations.

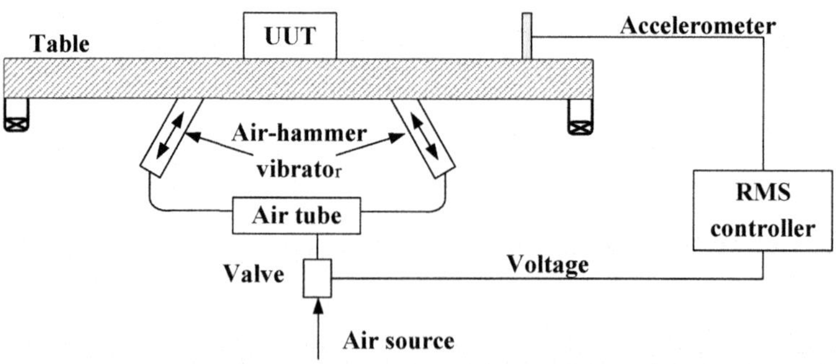

Fig. 3.17 Broadband pseudorandom vibration–temperature stress combined test platform

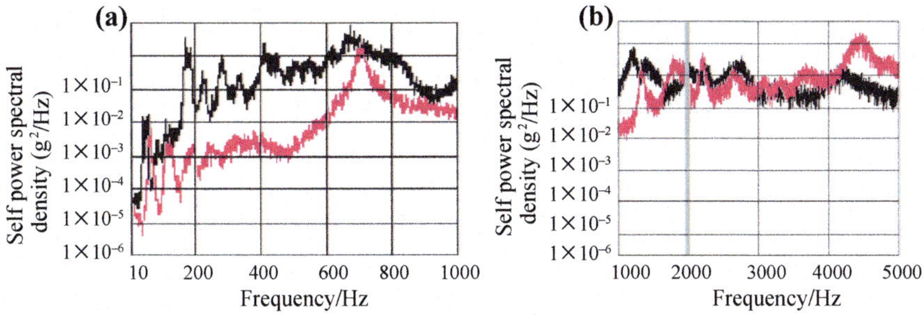

Fig. 3.18 Broadband pseudorandom vibration PSD

3.3 Verification Test Cautions

Many tests are required during a launcher's avionics development and production, as introduced in Sect. 2.2.1 and subsequently discussed. This section introduces only relevant items that are easily misunderstood or neglected.

3.3.1 Aging Tests

Screening tests include environmental stress screening and aging tests. Aging tests simulate service conditions and operate the product for some period to identify or eliminate early failures, maintaining low failure rates for delivered products. Aging is performed in an integrated assembly. However, for electromechanical equipment with limited service life, electronic components complete aging testing separately for relatively long periods and are then integrated for overall aging. Note that power-on time in screening is not included in the aging time.

Aging time depends on the expected product life, but is generally not more than 20% of product life. Where service life is evaluated by state change (switching on or off), aging should be conducted following the product specification or technical documents to determine state alteration times and frequency, generally not more than 5% of the expected life. For highly reliable applications, electronic component burn-in time is ~1000 h, while ground facility burn-in time is usually ~200 h.

Aging tests are divided into normal (25 °C) and high-temperature aging. High-temperature aging is preferred where the normal temperature burn-in time is too long. Temperature stress for high-temperature aging should not exceed the highest operational temperature, and high-temperature aging time is determined by engineering experience. However, aging time can also be accelerated as follows.

Based on Arrhenius model [13], the acceleration factor of high-temperature aging, A_F, can be expressed as

$$A_{\mathrm{F}} = \exp\left[\frac{E_a}{k}\left(\frac{1}{T_{\mathrm{u}}} - \frac{1}{T_{\mathrm{e}}}\right)\right], \tag{3.12}$$

where E_a is the activation energy (eV), and usually set as 0.6; k represents the Boltzmann constant (8.617×10^{-5} eV/K); T_{u} represents the absolute temperature when the equipment is used normally (K, degree Kelvin); and T_{e} represents the absolute temperature when the stress is acted on (K).

However, in practical cases, a simplified model is used to approximate accelerated aging time,

$$A_{\mathrm{F0}} = 2^{\left(\frac{T_{\mathrm{e}} - T_{\mathrm{u}}}{10}\right)} \tag{3.13}$$

and

$$t_{\mathrm{e}} = t_{\mathrm{u}}/A_{\mathrm{F0}}, \tag{3.14}$$

where A_{F0} is the acceleration factor of the simplified model, t_{e} is the aging time at stress temperature(s), and t_{u} is the aging time under normal service condition(s).

Taking 1000 h aging at room temperature (25 °C) as an example, the simplified model shows that aging time can be converted into 250 h at 45 °C.

Aging is continual through the operational period, and 4-h aging cycles are commonly adopted. Each cycle is a continuous power-on time, as described in the product specifications, and the test is repeated every 4 h. At the end of each cycle, power is left off for half an hour, which is not included in the aging time. Functional tests should be performed in every cycle, or every other cycle. Functional and performance (if possible) tests should both be performed during the final cycle.

Nominal electrical stress is assigned for 50% of aging time with upper and lower limits for 25% of the aging time, respectively. The distribution can be allocated proportionally in each cycle, or assigned by the number of the cycles, with constant stress over each cycle.

The aging time should be recalculated when failure occurs. It is essential to distinguish if the failure is due to an associated or non-associated fault. In the latter case, the test should continue after the fault is removed, and previous test time is valid. However, if it is an associated fault, improvements must be made with replacement components selected from within the same batch. The repaired UUT undertakes at least one test cycle to verify the improvement, and then continues the aging test. Aging time in this case is determined as follows.

1. If the failure occurred within the first half of the burn-in time, total burn-in time is accumulated in the time when the failure occurred. If continuous operation without failure reaches the required burn-in time, the test can be stopped.
2. If the failure occurred in the last half of the burn-in time, the test is considered complete when the continuous operation without failure after repair reaches at least 50% of the total burn-in time.

If multiple failures occurred, burn-in time should be determined in consultation with system designers (test supervisors).

3.3.2 Environmental Stress Screening

Environmental stress screening (ESS) of electronic devices is performed at the component, board, and equipment levels. The main test items include temperature cycle and random vibration. Temperature cycling methods are mature, so this section focuses on random vibration screening tests.

Product dampers must be removed in ESS, and the UUT rigidly connected with the vibration table. The aim of screening is to find defects introduced by manufacturing, and screening conditions should be independent of the practical UUT environment and what the UUT can endure. All UUTs are checked with the same criteria, with the applied vibration PSD, as shown in Fig. 3.19, based on extensive engineering practice with total RMS = 6.06 g.

If there is no suitable test assembly, the board can be installed in the chassis for screening. Vibration of the board's local parts may be magnified without the dampers, which may cause product damage or leave hidden dangers. The MIL-HDBK-2164A standard [14] requires that the spectral value be reduced to avoid potential fatigue or damage by peak stress when the resonance (Q > 10) on UUT parts occurs, commonly known as valley vibration [15]. Similarly, spectral magnitude should be increased for regions with significant attenuation. Therefore, trial vibrations should be performed in advance with sensors set where the board's maximum and minimum response locate, and the test profile shown in Fig. 3.19 adjusted accordingly. Two practical methods can be adopted to determine the screening conditions.

1. If the chassis response = 6.06 g and the maximum board response is less than 12 g, the condition is acceptable. This limitation (12 g) is based on experience, and can be adjusted depending on known UUT reliability.

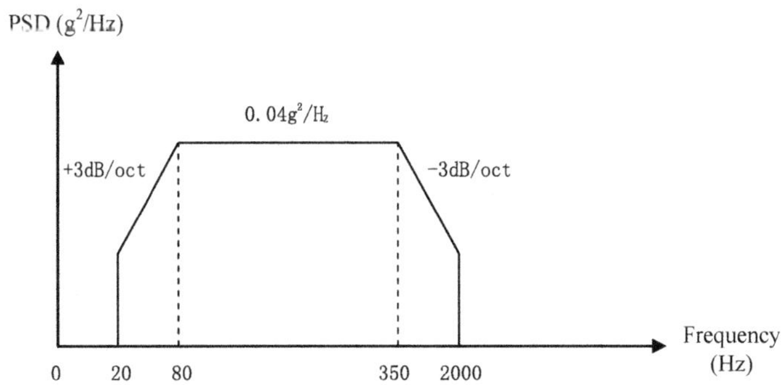

Fig. 3.19 Random vibration screening test PSD

2. If the chassis response = 6.06 g, but it is greatly magnified or weakened on the board, the spectrum should be adjusted to reduce or increase the PSD, respectively, to ensure the board monitor point response is approximately 6.06 g.

The first method avoids overstressing specific UUT components and boards, but may also cause less stress over others; whereas the second method balances the responses, but it is difficult to make each monitoring point response exactly 6.06 g due to various magnification effects on different parts of the board. Thus, slightly higher or lower responses are often acceptable.

It is important that trial vibrations are applied to prototypes rather than the formal devices, and the conditions determined by the trial should be formalized in the device specifications.

For integrated inertial devices, the body of inertial instruments and matched electronic circuits are installed in a chassis. Usually, the chassis has no dampers due to installation accuracy concerns, but the body is equipped with shock absorbers, commonly known as internal damping. Hence, the electronic circuits are rigidly connected with the chassis during both screening and acceptance tests, and circuit board vibration may be amplified. Therefore, vibration-sensitive components may be reinforced, the circuit board may be locally damped, or if necessary the chassis may be damped during screening to avoid negative effects from inappropriate screening conditions.

3.3.3 Other Test Requirements

3.3.3.1 EMC Test Requirements

The following test items and indexes are generally selected for LVs.

GJB151B [16]: CE102, CE106, CS101, CS103, CS104, CS105, CS106, CS114, CS115, CS116, RE102, and RS103, where CE106, CS103, CS104, and CS105 are only required for wireless devices.

Specific indexes are as follows:

1. Peak signal in CS106: $E = 200$ V, $t = 10 \pm 2$ s.
2. Maximum current in CS116 = 5.0 A.
3. Frequency range in RE102: 2–1000 MHz.

3.3.3.2 Low-Pressure Test Requirements

The low-pressure test initial condition = 101,325 Pa, with the lower limit depending on altitude. Generally, lower limit = 8×10^{-5} Pa for device installed in the last launcher stage. If the device does not meet these requirements, at least 1×10^{-3} Pa is required. Decompression rate is determined from the planned flight speed.

The UUT should be powered on during depressurization and low pressure. Depressurization shall ensure that duration time through the range for vacuum (10^{-3}–10^{-1} Pa) and micro (6.65–10^{-2} Pa) discharge is at least 10 min. For each of 133, 1.33, 0.1, and 10^{-3} Pa, the pressure shall be kept constant no less than 10 min and repeated at least three times. Both of these conditions are prone to instigate discharging.

If the thermal vacuum test is performed, heating flow is applied when pressure is $<10^{-3}$ Pa and stable.

3.3.3.3 Thermal Environment Test Requirements

The thermal environment test includes low- and high-temperature tests, and low temperatures are achieved with cryogenic tanks and pipes in-flight. The test is performed in a temperature test chamber. For UUTs that operate during takeoff, they start operating and then cooling begins in the test. When the temperature reaches the desired value, it should be held constant for three times the duration of minimum in-flight temperature. UUTs that are not operating during takeoff commence operation after the chamber reaches the desired temperature.

High temperatures in flight arise due to aerodynamic heating, jet scream heating, etc., and the heat flux test can also be performed if flux is known. The power-on time for the high-temperature test is the same as for the low-temperature test, and the temperature rise rate is 10 °C/min. If the temperature rise rate in-flight is greater, the equipment can be powered on and placed directly in a chamber that has already reached the specified high temperature.

3.3.3.4 Salt Fog Test Requirements

The salt fog test includes an accelerated laboratory test and natural salt fog test. The laboratory test references GJB150.11A [17], with spraying and drying circulating over 24 h each, i.e., each test cycle is 48 h, and four cycles are usually applied. If the UUT is installed at a coastal launch site, eight cycles (value obtained from previous experience) are preferable.

The natural salt fog test is more realistic and is recommended where possible. Experimental conditions and methods vary and are not discussed in detail here.

References

1. Blair JC, Ryan RS, Schutzenhofer LA. Lessons learned in engineering. NASA, NASA/CR-2001-216468
2. Ye YX (2010) Research and implementation of testing platform for embedded software. Beijing Jiaotong University, Beijing (in Chinese)
3. Dang JJ, Luo JJ, Wan YH (2010) Adaptability and navigation accuracy analysis of flexible gyro SIMU under vibration environment. J Projectiles Rockets Missiles Guidance 30(1):13–16 (in Chinese)
4. Xu QL, Deng ZL (2004) Test method of inertial instrument in LSINS. Aviat Metrol Measure Technol 24(6):25–27, 34 (in Chinese)
5. Wang ZW, Hou SM (2011) Rapid calibration method of strap-down inertial measurement unit. Missiles Space Veh (6):43–47 (in Chinese)
6. Song ZY (2012) Redundancy management and information fusion on meter-level for dual IMU. Manned Spaceflight 18(5):8–13 (in Chinese)
7. Feng WQ (2005) Spacecraft instrument environment reliability test. Spacecraft Environ Eng 22(10):249–252 (in Chinese)
8. Jin XS (2004) Spacecraft environmental test and space product quality and reliability guarantee. Chin Space Sci Technol 6:25–31 (in Chinese)
9. Zhou T, Hu HF, Chen WJ (2013) Research of reliability growth test based on flight test information as amended. Manned Spaceflight 19(4):11–16 (in Chinese)
10. Zhu XQ (2004) The application of reliability growth test in the development of the electric products of manned space launch vehicle. Missiles Space Veh (1): 61–66 (in Chinese)
11. Wang FF (2002) Modified highly accelerated life test for aerospace electronics. In: The Inter-society Conference on Thermal and Thermomechanical Phenomena in Electronic Systems (ITHERM), San Diego, May 30–June 1, 2002
12. Fan RX, Liu JZ, Xu WB et al (2014) Engineering practice of reliability strengthening test for newly developed rocket. In: 3rd manned spaceflight academic conference, Chengdu, Nov 13–15, 2014 (in Chinese)
13. Kececioglu D, Jacks JA (1984) The Arrhenius, eyring, inverse power law and combination models in accelerated testing. Reliab Eng 8(1):1–9
14. MIL-HDBK-2164A (1996) Environmental stress screening process for electronic equipment
15. Zhu YC (2007) Applications and developments of environmental stress screening (ESS). Spacecraft Environ Eng 24(4):191–197 (in Chinese)
16. GJB 151B (2013) Electromagnetic emission and susceptibility requirements and measurements for military equipment and subsystems (National military standard, in Chinese)
17. GJB150.11A (2009) Laboratory environmental test methods for military materiel—Part 11: salt fog test (National military standard, in Chinese)

Chapter 4
System Level Testing

Launch vehicle tests include many important activities, particularly during the development phase. However, when the rocket begins launch services, system level tests focus on avionics system integration after assembly, mainly by simulating flight processes, also known as general check in China. The general check is performed hierarchically. The control (or GNC) subsystem is checked first, and then other systems, such as the telemetry system, etc. The shutdown time of every stage may be controlled around the theoretical value by modeling the shutdown equation.

In the general check, flight control system electrical performance is closely examined, including hardware and software function and compatibility, coordination between different devices, etc. Coordination between GS, TLCS, and onboard flight control systems are assessed, and test function, information display and monitoring, and TLCS launch processes are also checked. The electromagnetic compatibility between GS and OBS is verified. And depending on the requirements, the typical redundant design and emergency processing can also be evaluated using system level testing.

The pre-launch check is also important. Since the pyrotechnics have been installed and propellant has been filled, the ignition and cutoff commands should not be executed in the general check. The pre-launch check can also be divided into three categories:

- single functional tests, e.g., ignition and cutoff circuits inspection, safety monitoring for sequential commands, aiming function tests, perpendicularity adjustment, etc.;
- functional survey tests combining several subsystem test items; and
- real-time monitoring, e.g., sampling important signals and displaying outcomes on the terminals in real-time, including power supply (voltage, current, frequency, etc.), electric leakage status, temperature, hydraulic control pressures, safety-related indicators, etc.

Z. Song, *The Test and Launch Control Technology for Launch Vehicles*,
https://doi.org/10.1007/978-981-10-8712-7_4

However, early launcher development was usually independent, so system test activities vary in different countries, as briefly described Chap. 1. Comparing the US and China schemes, NASA employs closed-loop integration testing to verify electrical systems (including the GNC algorithm); whereas in China, aerospace industry departments are often both the sponsor and contractor for testing, and tests are divided into integration tests, to assess avionics system coordination; and semi-physical simulation tests, to verify the GNC algorithm. The former is mainly open loop, and can be implemented in the assembly workshop and launch site; the latter is mainly performed in the design department (see Chap. 5).

This chapter focuses on the test scheme, items, and methods for the Long March launch vehicle (LM LV) electrical systems. Test data analysis is very important, and manual evaluation cannot meet the demands of increasing test data volume. Therefore, automatic analysis has been developed, as discussed in Chap. 7.

4.1 Subsystem Testing

Testing cannot be covered by a single standard. The US Military adopts unit, subsystem, and vehicle tests for LV, upper stages, and spacecraft. Unit testing is similar to equipment level testing introduced in Chap. 3. Subsystem and vehicle testing are both called system level testing in China, but with different scopes. Although test classifications and requirements are described in the relevant standards, e.g., MIL standards with the purposes clarified to ensure the rocket design meets the requirements, there are no guidelines on how to perform the tests. Rockets vary, hence test methods vary accordingly, and are deeply affected by the experience.

Electrical system testing is required to be convenient to perform in the assembly factory and launch site in China, focusing largely on coordination between different subsystems, where the connection state should be kept as close to in-flight state as possible. Testing is progressively integrated, starting with subsystem testing. However, simulation tests usually change the connection state of the whole system and many simulators are involved, which is unsuitable for utilizing outside design departments or contractors.

This section introduces general system test methods for onboard control systems [1–3], which are usually divided into single function, subsystem tests, and general (also knows flight simulation) check. Test methods for other systems are the same as for the control system [4]. The test system should also provide real-time monitoring of key signals.

Single function tests are applied for test items that are difficult to incorporate into automated test processes, or as preparation before automatic tests. They have little correlation with each other, and are generally only performed during initial testing, mainly including ground power supply start and stop, ground and onboard power supply match, exhaust signals inspection, GS and OBS connector separation, sequential control circuit safety check, reset examination, and unloaded rotating table testing. These tests are not discussed in detail here.

Subsystem tests are functional static tests that measure the response or output of the device or subsystem under specific stimulations. The definition of the subsystem is somewhat relative: the overall control system can itself be considered a subsystem, but generally, it refers to a smaller subsystem composed of several devices. The combination of subsystem tests should cover all control system devices.

4.1.1 Test Items

Test facilities are integrated through the standard bus, introduced in Sect. 2.3.3. Subsystem tests are usually divided into several items, tested independently or in combination. Test stimulations and ideal responses are determined, allowing automatic analysis of test outcomes. The various test items are generally divided into the following categories.

4.1.1.1 Attitude Control System Tests

The overall attitude control system is usually tested as separate stages, i.e., boosters, first core stage, second core stage, etc., as shown in Fig. 4.1.

The main purpose of attitude control tests is to check the match between attitude control equipment, particularly static gain and polarity. User-defined test software is installed in the OBC, and control law parameters can be different from those in flight.

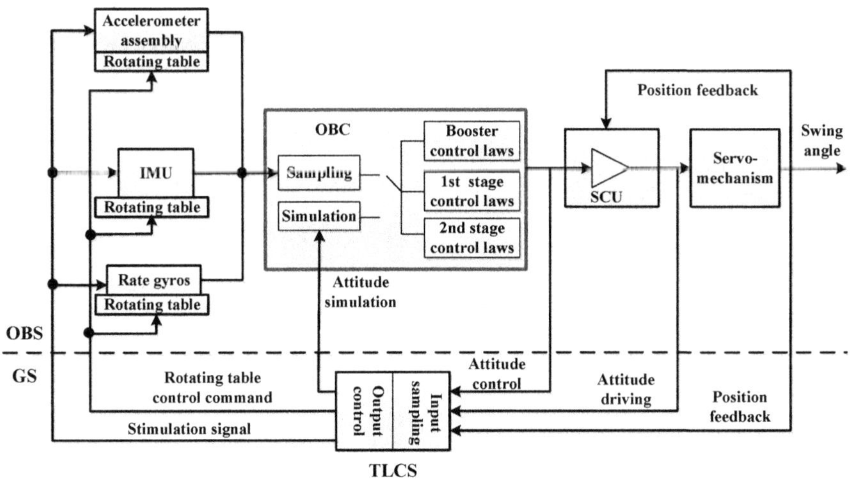

Fig. 4.1 Attitude control system test block diagram

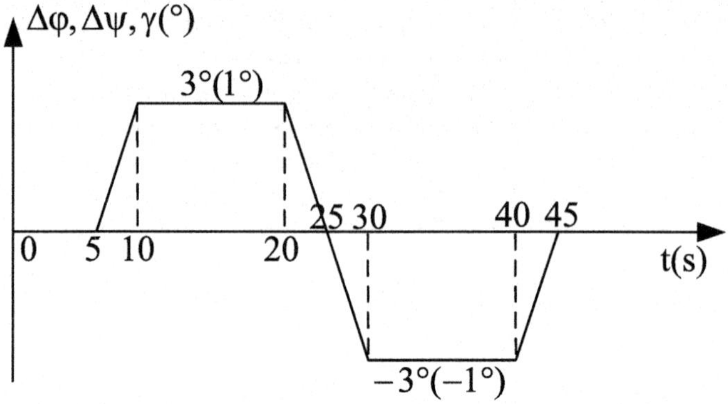

Fig. 4.2 Typical attitude control signal during testing

Testing checks the loop: inertial device output, OBC input, control laws, OBC output, servo driver outputs, and servo mechanism feedback. Inertial device outputs can be realized by rotating the table, and mechanical gyro and accelerometer outputs can be stimulated by imposing constant current to their torque drivers. If these approaches are not available, simulated inertial signals can be sent to the OBC directly, or the OBC directly sends simulated signals to the servo drivers. The test system samples OBC and SCU (or driver) outputs, and servo mechanism feedback, allowing the loop static gain to be calculated and verified.

If the inertial device has been installed onboard rather than onto a rotating platform, inertial signals can only be produced by torquer excitation, and the servo mechanism would swing the engines. If the launcher is erect, this swing may cause oscillation, which could shake the launcher in some specific circumstances, with significant safety implications. Therefore, a filter network is added to the control laws to eliminate oscillations during vertical testing.

Polarity check is also a test focus. A stable control signal is generated to check the engine swing direction by servo mechanisms, and a typical attitude control signal is shown in Fig. 4.2. The control signal can instruct pitch, yaw, or roll separately or simultaneously, and static gains should be checked in different flight stages to ensure test coverage.

4.1.1.2 Inertial Device Tests

Test items include the checks for IMU, accelerator assembly, and rate gyros. For optical gyros, it is difficult to stimulate outputs by torquer excitation, tests are combined with sensing the Earth's rotation. Depending on how the IMU is set up, Earth rotation angular rate and local gravitational acceleration can be converted into ideal gyro and accelerometer measurements along orthogonal directions, and these ideal values are compared with actual inertial device outputs. This is called rational judgment. In contrast, rotational angular rate and local gravitational acceleration can be calculated from the inertial device outputs, and compared with the theoretical value, which is called comprehensive judgment.

Comprehensive Judgment

Comprehensive judgment is not used to distinguish individual meter outputs, but rather for overall IMU evaluation. The methods are as follows:

$$\Delta\omega = \left| 15 - \sqrt{\left(\frac{K_{gx} * N_{GX}}{t} - D_{0x}\right)^2 + \left(\frac{K_{gy} * N_{GY}}{t} - D_{0y}\right)^2 + \left(\frac{K_{gz} * N_{GZ}}{t} - D_{0z}\right)^2} \right| \tag{4.1}$$

where N_{GX}, N_{GY}, and N_{GZ} represent the cumulative number of gyro pulses in t seconds, respectively; $K_{gx}, K_{gy}, K_{gz}, D_{0x}, D_{0y}$, and D_{0z} represent the pulse equivalent and zero-term drift of gyros, respectively; $\Delta\omega$ is the difference between the Earth's rotational rate (15″/s) and the RMS of three orthogonal gyro outputs, used to check if the gyros work well.

$$\Delta G = \left| 1 - \sqrt{\left(\frac{N_{AX}}{t * K_{ax}} - K_{0x}\right)^2 + \left(\frac{N_{AY}}{t * K_{ay}} - K_{0y}\right)^2 + \left(\frac{N_{AZ}}{t * K_{az}} - K_{0z}\right)^2} \right|. \tag{4.2}$$

where N_{AX}, N_{AY}, and N_{AZ} represent the cumulative number of accelerometer pulses in t seconds, respectively; $K_{ax}, K_{ay}, K_{az}, K_{0x}, K_{0y}$, and K_{0z} represent the pulse equivalent and zero-term drift of accelerometers, respectively; ΔG is the difference between the local gravity acceleration ($1g_0$) and the RMS of three orthogonal accelerometer outputs, used to check if the accelerometers work well.

For $\Delta\omega$ and ΔG, suitable thresholds are set to judge if the deviations exceed them.

Decomposition Judgment

The IMU is always in the vertical state during testing. Hence, acceleration (1g) is perpendicular to the local level, and the angular rate is Earth's rotation projected on the three axes. Errors between the ideal and actual measurements depend on IMU calibration parameters, leveling, and aiming errors.

For example, if launch direction is A (°), latitude is B (°), and Earth rotational rate is ω_e (°/h), then the rotation components along the pitch, yaw, and roll axis in IMU coordinate system are

$$\begin{cases} \omega_z = \omega_e \times \cos B \times (-\sin A) \\ \omega_y = -\omega_e \times \cos B \times \cos A \\ \omega_x = \omega_e \times \sin B \end{cases}, \tag{4.3}$$

where ω_x,ω_y, and ω_z represent the rotation components along the roll, yaw, and pitch axis, respectively; And then the theoretical pulses of each gyro in t seconds N_{gx}, N_{gy}, and N_{gz}, are

$$\begin{cases} N_{gx} = \frac{(D_{0x} + \omega_x + D_{xy}\omega_y + D_{xz}\omega_z)}{K_{gx}} \times t \\ N_{gy} = \frac{(D_{0y} + \omega_y + D_{yx}\omega_x + D_{yz}\omega_z)}{K_{gy}} \times t \\ N_{gz} = \frac{(D_{0z} + \omega_z + D_{zx}\omega_x + D_{zy}\omega_y)}{K_{gz}} \times t, \end{cases} \tag{4.4}$$

where $D_{xy}, D_{xz}, D_{yx}, D_{yz}, D_{zx}$, and D_{zy} represent the linear term cross coefficients between different gyro measurement axes, respectively.

In the calculation of theoretical pulses of each accelerometer in t seconds, N_{ax}, N_{ay}, and N_{az}, the cross coefficient of the linear term is approximately 0, and the quadric term is also 0, to minimize the computing overhead:

$$\begin{cases} N_{ax} = (K_{0x} + \dot{W}_x) \times K_{ax} \times t \\ N_{ay} = (K_{0y} + \dot{W}_y) \times K_{ay} \times t, \\ N_{az} = (K_{0z} + \dot{W}_z) \times K_{az} \times t \end{cases} \tag{4.5}$$

where $\dot{W}_x, \dot{W}_y, \dot{W}_z$ are the apparent acceleration (corresponding to local gravity acceleration) components of the three IMU acceleration measurement axes, respectively.

As discussed above, torquer excitation can also be applied to mechanical gyroscopes and accelerometers. For the accelerometer,

$$N_{ax} = (K_{0x} + \dot{W}_x + IK_{ig}) \times K_{ax} \times t, \tag{4.6}$$

where I is torquer excitation current (mA), K_{ig} is the torquer factor (g/mA).

Rotating Platform Tests

Rotating platform tests examine IMU function and are difficult to assess accuracy. The platform rotates based on the platform controller command, with the rotating speed sensed by the IMU and checked against the controller. When the platform rotates to a new fixed position, the local gravity acceleration can be decomposed for each IMU measurement axis, and compared with the accelerometer outputs.

Accelerometer assembly tests are similar. When installed vertically, transverse and normal components are almost zero, hence torquer excitation is preferable for testing. If installed on a platform, the platform can turn so the meters can sense changing gravity components. For rate gyros, the platform rotational speed or the Earth angular rate can be measured and checked.

Consider the pitch rate gyro as an example,

$$\dot{\varphi} = N_{\varphi} \times K_{\varphi}. \tag{4.7}$$

where $\dot{\varphi}$ is the rotational rate, N_{φ} is the rate gyroscope digital output (pulse), K_{φ} is the rate gyroscope equivalent.

The rate gyro's stability over some defined period, e.g., 2 min,

$$\sigma_{\varphi} = \sqrt{\frac{\sum_{i=1}^{n} (\dot{\varphi}_i - \bar{\dot{\varphi}}_i)^2}{n-1}}, \tag{4.8}$$

where suffix i represents the ith sampling of rate gyroscope output, $\bar{\dot{\varphi}}_i$ represent the average value of all n sampling data, σ_{φ} is the standard deviation of these data. The deviation depends on the rate gyro accuracy and the installation errors.

For redundant design, if the relative installation errors between these IMUs can be measured accurately, then one IMU can act as the reference for the others. Thus, in-flight FDI can also be employed for ground tests.

4.1.1.3 Sequential Circuit Tests

The OBC transmits a command to the ICUs, and ICU timing and sequential commands are sampled by TLCS to verify if they meet the specifications. Sequential commands are instructions sent depending on a pre-designated time sequence, and have a short duration, e.g., 50–300 ms, and strict order. In some LVs, pressure control is also done by ICU, so the test provides a comprehensive check for discrete control signals. Pressure sensor signals must be simulated as inputs for pressure control test, and varied, to increase the test coverage, i.e., a fixed simulated sensor value and fixed controller response are insufficient. Figure 4.3 shows a typical test module.

For these discrete control signals, the magnitude is the same as the power supply voltage. The starting time and duration of each instruction should be recorded, and the instruction starts at various times for different test cases, so the measurement is triggered in interrupt mode.

The module includes bus interface control logic, configuration registers, timers, signal jump detection, time and state registers, memory, input optocouplers, filters, etc. Filters minimize noise signal caused by electromagnetic relay contact bounce, and the filter width can be set in advance. Optical isolation provides electrical and other disturbance protection. When input signal jumping (from low to high voltage to start, or from high to low voltage to clear) is detected, the jump detection circuit records the state and current time to a set of registers, setting the state bit and triggering an interrupt request to the test bus controller. Significant memory capacity is required to record every jump. Application software receives the recorded information in interrupt mode, or by polling the registers.

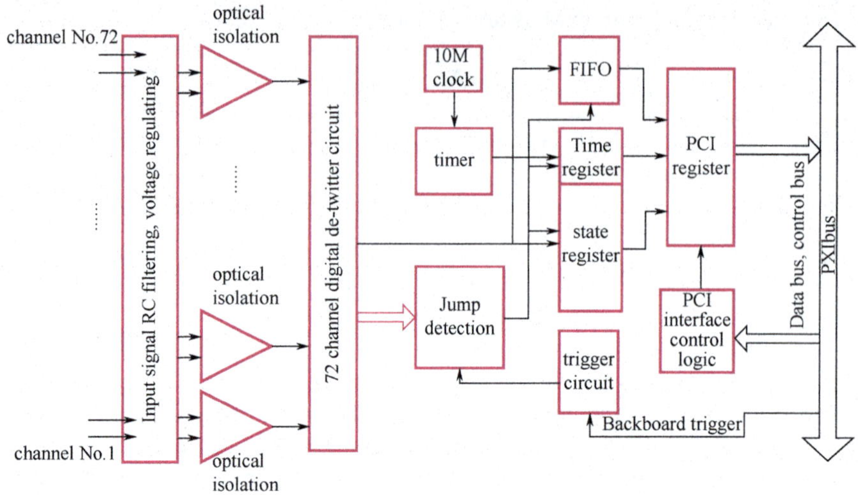

Fig. 4.3 Sequential signal test module block diagram

Table 4.1 Shutdown mode combinations during sequential command check (example)

State	Booster 1, 3	Booster 2, 4	First core	Second core main engine	Second swing engines
1	Exhaustion	Exhaustion	Range	Semi-major axis	Guidance cutoff
2	/	/	Range	Semi-major axis	Exhaustion + guidance cutoff
3	Exhaustion	Exhaustion	Exhaustion	Exhaustion	Exhaustion + guidance cutoff
4	Exhaustion	/	Exhaustion	Semi-major axis	Exhaustion + guidance cutoff

Sequential commands tested include rocket ignition, shutdown, stage separation, and other functions, each of which has normal and backup states with activation permission constraints. All conditions should be checked. Table 4.1 shows an example of different test cases, where the combinations of varied shutdown modes are checked. Actual in-flight combinations are somewhat more comprehensive than this short list, and system level testing only checks several typical modes.

4.1.1.4 Satellite-Integrated Navigation Interface Tests

Figure 4.4 shows the satellite-integrated navigation interface check structure. The main inspection parameters include

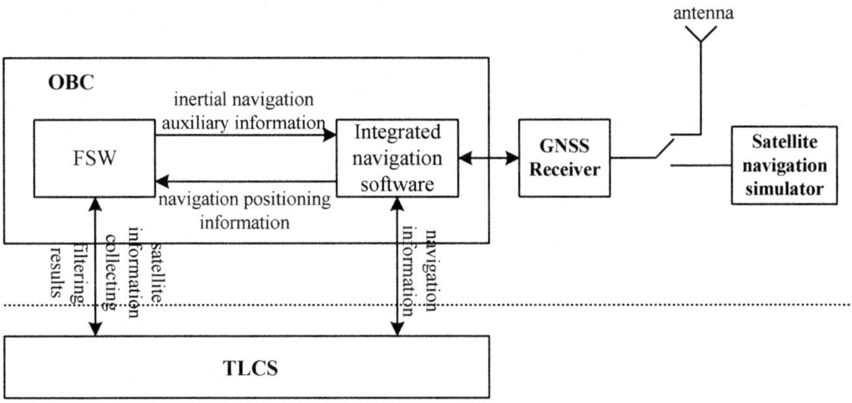

Fig. 4.4 Integrated navigation interface check diagram

1. Time, velocity (*Vx, Vy, Vz*), and position (*Px, Py, Pz*);
2. Receiver positioning mode;
3. Receiver status and error code;
4. Position dilution of precision.

When the satellite simulator is adopted, the flight software also transmits the filter results to the test system.

The tests check the receiver, receiver and OBC communication, navigation and flight control software interface, etc.

4.1.2 *Redundancy Testing*

The redundant design is widely used in launcher control systems to improve reliability, but also increases the difficulty in exposing the fault during ground testing due to the redundancy. Therefore, testing must identify failures in each redundant module to ensure that they operate normally prior to flight. Redundant tests are required for unit and system level testing.

Redundant design of the LM LV control system is divided into a TMR-based voting system, fault diagnosis-based redundancy management, failure absorption-based fault-tolerant design, parameter margin-based and component level redundant design. These reliability design measures are verified in different development phases. Although they are examined in detail during equipment level testing, they should also be checked at the system level if convenient.

The TMR-based voting mechanism can be tested as follows.

1. If the redundant part can be powered on separately, breaking the power supply of one part will be equivalent to simulating a part fault.

2. If the embedded software is installed via communication rather than solidified, a part fault can be simulated by not uploading the software.

Fault diagnosis-based redundancy management is mainly used for inertial device FDI. IMU differences can be simulated in the system test by torquer excitation and the FDIR algorithm examined. Parameter margin-based redundant design is mainly checked by simulation or identification tests. Component level redundant design mainly adopts parallel, serial–parallel, or parallel–serial connections, and is usually tested at the equipment level. If DFT is adopted, redundancy checks can also be made at the system level.

The sections consider test schemes for failure absorption-based fault-tolerant techniques, and two examples are presented to illustrate the test schemes.

4.1.2.1 Digital to Analog Output Fault Absorption Test

Figure 4.5 shows the fault-tolerant architecture. Three identical limiting resistors on the output end of the operational amplifier (OP-AMP) are in the feedback loop to adjust and balance output differences when the TMR circuits operate normally. The feedback resistor provides current-limiting protection and balances the bias between the OP-AMP positive and negative ends. The three protection resistors on the redundant output end protect the inner interface circuit from damage by external abnormal high voltage.

Absorption is achieved by current-limiting protection, where the OP-AMP maximum output current = 30–40 mA. When a D/A fails, its output circuit is equivalent to the load of the other two. Since the OP-AMP maximum input/output current is half that of the other two (determined by current-limiting resistors), the

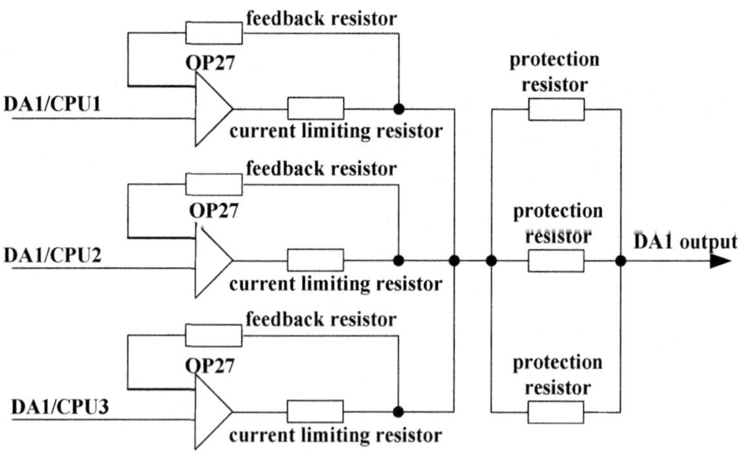

Fig. 4.5 Fault absorption function of D/A output

equivalent load will not affect the normal output. Therefore, the scenario where two OP-AMPs output the same voltage while the other is different, even with opposite polarity and the same magnitude, can be simulated to check the effect.

4.1.2.2 Servo Valve Failure Absorption Test

Failure absorption technology [5] is also adopted in the servo mechanism closed-loop control. This control subsystem comprises a servo driver and servo mechanism. The servo valve drivers, preamplifier, and actuator position sensors are TMR configured [6, 7], as shown in Fig. 4.6. The servo drivers output current to the servo valve coils, the valves drive the actuator, and the actuator position is sensed and fed back to the drivers, creating a closed loop.

Failure absorption principle details can be found in [6]. We describe a test scheme to verify fault absorption by externally applying a reverse current in system level testing, as shown in Fig. 4.7.

A testing voltage, U_t, is imposed on the resistance, R_t, creating a reverse current, I_{rc}. The TLCS controls reverse current switching, the current flows through one servo control loop. Since the feedback resistance, R_f, is very large, the current may be approximated as

$$I_{rc} = \frac{U_t - U_1}{R_t} \tag{4.9}$$

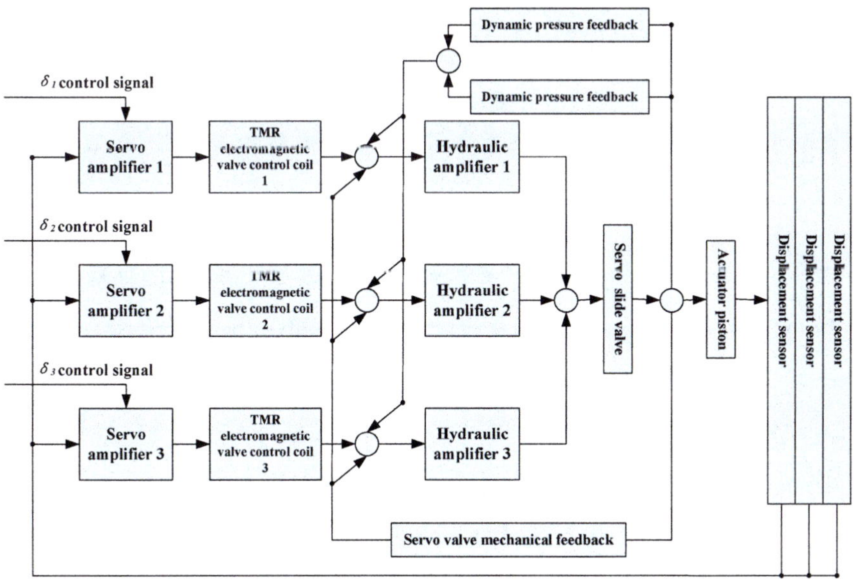

Fig. 4.6 Servo control loop block diagram

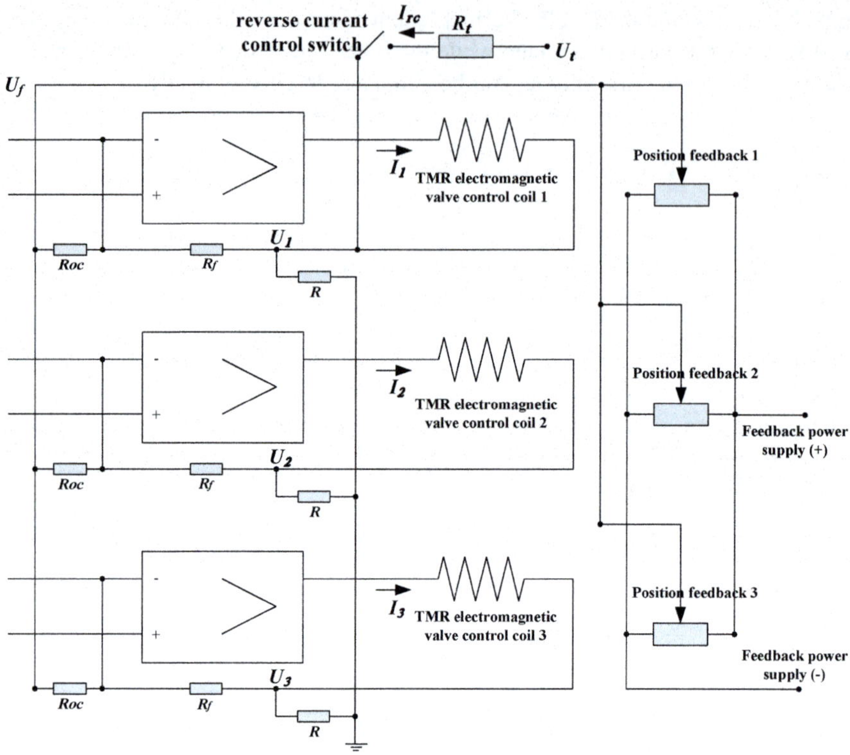

Fig. 4.7 Reverse current failure simulation test

$$I_1 = \frac{U_1}{R} - \frac{U_t - U_1}{R_t} \tag{4.10}$$

$$I_2 = \frac{U_2}{R} \tag{4.11}$$

$$I_3 = \frac{U_3}{R}. \tag{4.12}$$

With feedback, the sum of the current in the three valve control coils is zero, i.e.,

$$\frac{U_1}{R} - \frac{U_t - U_1}{R_t} + \frac{U_2}{R} + \frac{U_3}{R} = 0. \tag{4.13}$$

Thus,

$$U_t = (U_1 + U_2 + U_3) \times \frac{R_t}{R} + U_1. \tag{4.14}$$

For

$$U_1 = U_2 = U_3 = \frac{U_f \times R_f}{R_{oc}}, \tag{4.15}$$

then

$$U_1 = U_2 = U_3 = \frac{U_t \times R}{3 \times R_t + R} \tag{4.16}$$

and we can define the balance current, e, as

$$e = \frac{U_t}{R_t} - U_1 \times \left(\frac{1}{R} + \frac{1}{R_t}\right) - \frac{U_2}{R} - \frac{U_3}{R}. \tag{4.17}$$

If fault absorption is successful, the actuator is maintained at the original position and $e \approx 0$. Since each redundant part of the servo control loop has slightly different parameters, e is nonzero but small. When e is large or the servo mechanism drifts significantly from the original position, fault absorption has failed.

4.1.3 Pre-launch Check

The pre-launch check is the final examination before launch, and includes some single functional and subsystem tests, generally as follows:

1. Check pyrotechnic and electromagnetic valve control loop resistances to ensure each loop is closed and properly connected, with no missed or incorrect connection.
2. Check the exhausted signal circuit. Powering the exhaust sensor will generate an exhausted signal, usually a discrete switch-on signal, because the sensor has not been immersed in the propellant yet and its output is equivalent to the exhausted state.
3. Check sequential control circuit to ensure initial safe states, and avoid false starts when powered on.
4. Rollcall check for bus terminals, and collect each terminal self-test result.
5. Check all inertial devices.
6. Check the satellite-integrated navigation subsystem.
7. Check the attitude control subsystem.
8. Check the aiming subsystem.
9. Check LV verticality.

Checks 5–7 are the same as those introduced in Sect. 4.1.1. Some of these check items are described below.

4.1.3.1 Pyrotechnics and Valve Control Circuit Tests

Launch vehicles include many pyrotechnics and valves, particularly for large LVs. These products have significant safety concerns, and pyrotechnics are usually only installed in the final phase before launch. Since they are installed throughout the LV, a major focus of the pre-launch check is to ensure all pyrotechnics and valves are correctly connected. Methods to effectively and safely check connection correctness has attracted considerable attention.

This check is manual for the LM LV, using a special test facility to measure loop resistances where the pyrotechnics and valves are included. Circuits are open state when unconnected, and their resistance is inconsistent with the expected value when incorrectly connected. However, if many pyrotechnics are parallel connected, particularly where a single sequential command detonates multiple pyrotechnics simultaneously, one (or potentially more) missed connections causes no significant change in overall resistance, and may cause an incorrect judgment. On the other hand, manual testing is inefficient and takes significant operation time.

Thus, automated methods are essential [8]. This section introduces several automated methods, including for electromagnetic valves testing. Chapter 7 presents another onboard safety protection and protection release scheme for these circuits, based on calculating the loop resistance using constant test current and voltage.

The DFT approach is applied for onboard devices before testing, where there are enough test ends. Table 4.2 shows the definition of the resistor box, and the following sections follow this configuration. The test ends are both set up on the resistor box, including all contacts assigned for short-circuit protection.

Current-Limiting Resistance Test

Figure 4.8 shows the current-limiting resistance test circuit. Consider a dual-bridge pyrotechnic, H1 and H2, for example, connected in parallel, with diodes to prevent mutual coupling between the test signal and provide isolation. Diodes are also connected in parallel to prevent open fault of one component. Pyrotechnics should

Table 4.2 Resistor box connection point distribution

Serial number	Connector identifier	Function
1	X1	Sequential command output signal
2	X2	Power supply negative bus
3	X3	Pyrotechnics (electromagnetic valve) positive end
4	X4	Pyrotechnics (electromagnetic valve) negative end
5	X5	Short-circuit protection/loop resistance test end
6	X6	Sequential signal self sampling end
7	X7	Telemetry sampling end

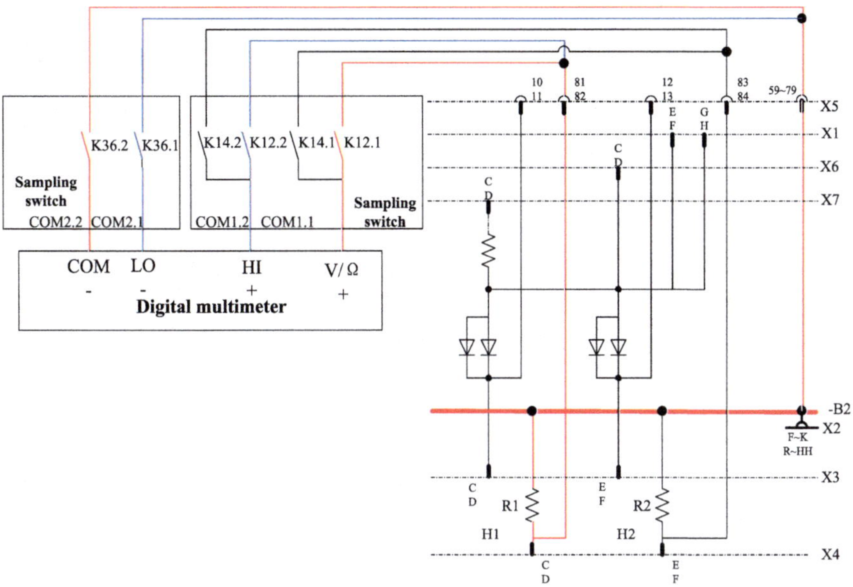

Fig. 4.8 Current-limiting resistance test principle diagram

be connected between the X3 and X4 connectors, but they are disconnected in this test. $R1$ and $R2$ are current-limiting resistors to be tested.

The test facility consists of a portable programmable digital multimeter and corresponding sampling switch, and a four-wire resistance measurement is adopted (the number of contact points and switches shown are just an example).

The red line in Fig. 4.8 represents the constant current path. The current flows from switch K12.1, through resistor $R1$ to the −B2 end, and then to the common end via switch K36.2. The blue line represents the voltage to be tested, sampling both ends of the resistor to provide the resistance. The measured four-wire resistance is the resistance between the ends of the sampling point, that is, the resistance from X5/81, 82 to X5/59–70. The resistance is the sum of R1 and wire resistance in the resistor box (wire resistance testing is introduced in the following section).

Wire Resistance Test

To test the wire resistance in the resistor box, a short-circuit is connected where the pyrotechnics would connect, and then follow the circuit shown in Fig. 4.9. If X3/C, D and X4/C,D are shorted, wire resistance in the resistor box can be measured. The diodes prevent the test constant current from passing through other channels. The measured resistance is subtracted from that measured in Current-Limiting Resistance Test to obtain the accurate current-limiting resistance, $R1$.

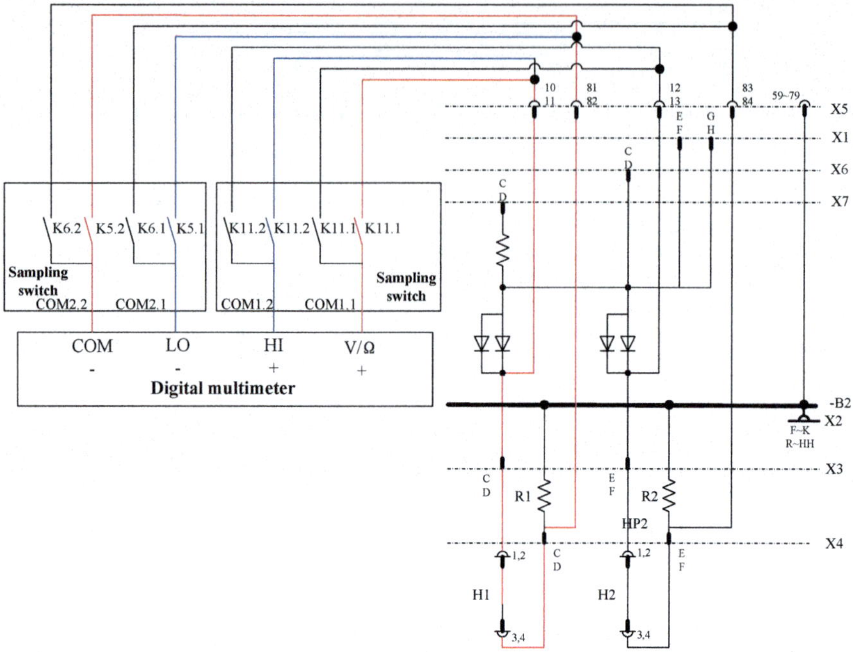

Fig. 4.9 Onboard wire resistance test principle diagram

Loop Resistance Test

After the pyrotechnics are connected, the total resistance of each control circuit can be measured following the loop shown in Fig. 4.10, which subtracts the current-limiting resistance to obtain the pyrotechnics resistance.

Electromagnetic Valve Control Loop Resistance Test

Electromagnetic valves are used for engine and pressure control. Valve resistance can be measured similarly to the loop resistance test, as shown in Fig. 4.11, where the valve is connected rather than pyrotechnics.

Electromagnetic Valve Operational Characteristics Test

Some valves must be checked that they can be switched on/off normally. If it is powered on/off, its current can be monitored to judge its operational state. Usually, the current is converted to voltage to be tested. A small resistor is connected to the valve control loop, and the voltage across it provides the valve current, as shown in Fig. 4.12.

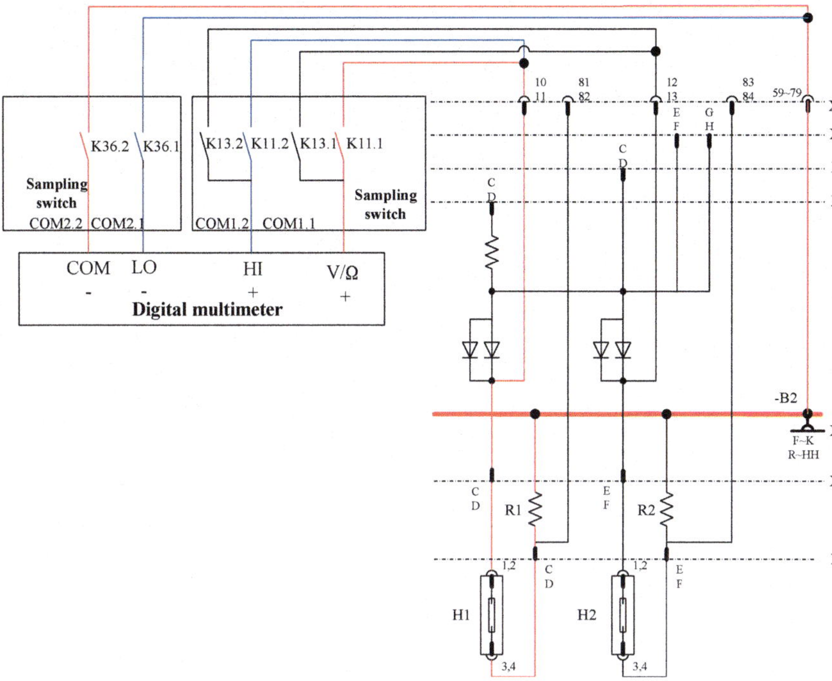

Fig. 4.10 Loop resistance test principle diagram

The sampling resistor, *R*, is connected in series with the valve, and the valve is powered on by the battery through the switch. A digital multimeter measures the voltage across *R*, divided by R to provide the valve current.

4.1.3.2 Sequential Control Circuit Safety Check

The initial state is very important for the sequential control system, and serious accidents could occur if an inappropriate initial state occurs when the system powers on. Thus, checking the system initial state is essential.

The desired initial state can be achieved by ordering the power-on sequence. First, the control circuit should be powered, with all output circuits remaining uncharged. Thus, even were some commands to be sent due to the wrong initial state, it will be unable to produce any effect. Then, the control circuit is reset so all commands are in invalid states. Finally, the output circuit is powered up.

The reverse order should be followed when powering off, ensuring output circuits are powered off first, and then the control circuit.

All sequential control circuit states must be checked after power-on. This is achieved following two methods, depending on the relay type, electromagnetic or solid state.

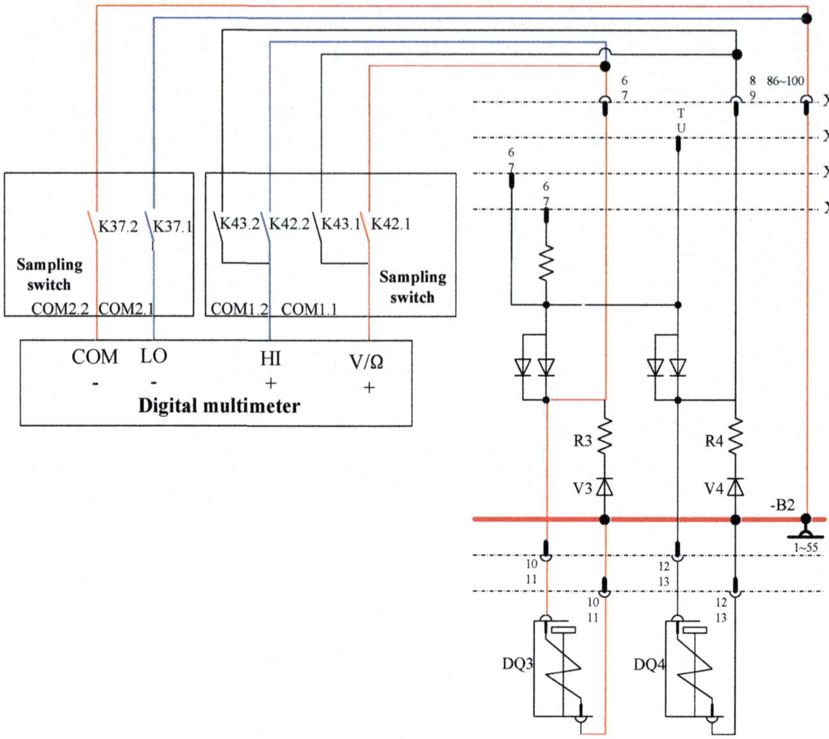

Fig. 4.11 Electromagnetic valve control loop resistance test principle diagram

Figure 4.13 shows the process for electromagnetic relays. Switches KT1–KT3 are controlled by TLCS. If one pyrotechnics control loop is triggered, e.g., K1 is on, then a loop is created: ground power supply (positive) → KT1 → Optocoupler → KT3 → X1TC/a → pyrotechnics power bus (positive) → 1K1, 2K1 contacts → pyrotechnics → pyrotechnics power bus (negative) → X1TC/b → KT2 → ground power (negative). Thus, the optocoupler is triggered, and a safety alarm is raised. Any channel error will cause an alarm.

Since the pyrotechnics is in the loop, the check must ensure safety. Current where one control loop is wrongly triggered should be restricted so it cannot trigger the pyrotechnics explosion, hence R_1 and R_t are current-limiting resistors; Moreover, KT1–KT3 contacts are connected in series to ensure that even if one is incorrectly connected, the loop will still be reliably off after the check.

Figure 4.14 shows the safety check process for solid-state relays, which is somewhat similar to the electromagnetic relay case. However, due to the solid-state relay leakage current, a threshold is required to distinguish between leakage and short-circuit current. Therefore, current monitoring is converted to voltage inspection, as shown in the figure.

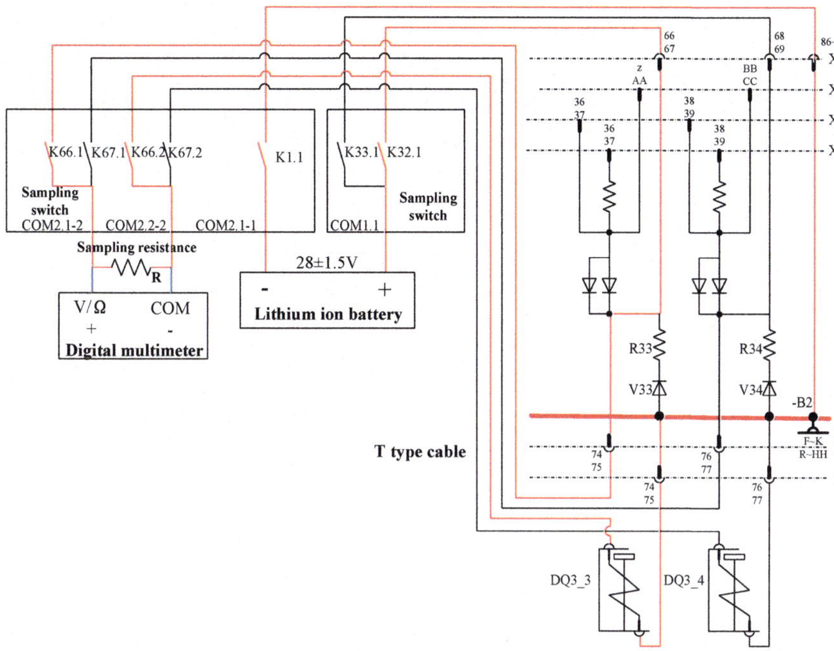

Fig. 4.12 Electromagnetic valve operational characteristics test principle diagram

For normal solid-state relays, if the leakage current is I, then the maximum total leakage flow $I_{tl} = 2 \times n \times I$, where n is the number of sequential control loops. If $n = 60$, and $I = 10$ uA, then the voltage across the 4 kΩ resistor is $V_{tl} = 1.2 \text{ mA} \times 4 \times 10^3 = 4.8$ V. When there is a short-circuit in the solid-state relay, ignoring the optocoupler current limit and bypass resistance, the test end voltage (V_{tl}) should be close to the supply voltage, i.e., 28 V. Therefore, if the measured voltage is less than some threshold, it is caused by the leakage current; whereas if it exceeds the threshold, a short-circuit has occurred.

The 4 kΩ resistance is selected to limit the current below 10 mA when a solid-state relay short-circuits occur, to ensure it will not trigger the pyrotechnics explosion. Parasitic capacitance at the output terminal of the solid-state relay can also be pre-charged during the check. After power switches to the onboard power supply, the resistor should be shorted by the switch.

4.1.3.3 Perpendicularity Adjustment Test

The aim of perpendicularity adjustment is to ensure that first-stage engine thrust is coincident with the local vertical line as much as possible, reducing lateral drift during liftoff. It is also used to measure the levelness between the base plane where the IMU is installed and the launcher base. Differences sometimes occur due to

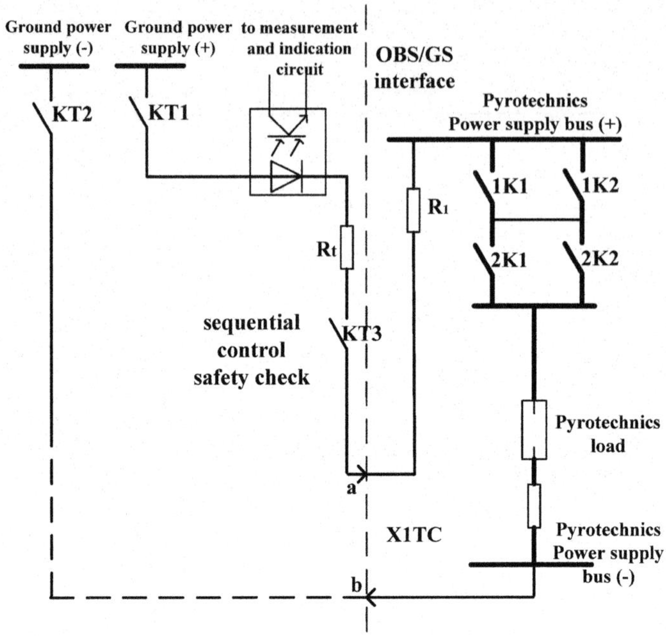

Fig. 4.13 Sequential control safety check for electromagnetic relay circuits

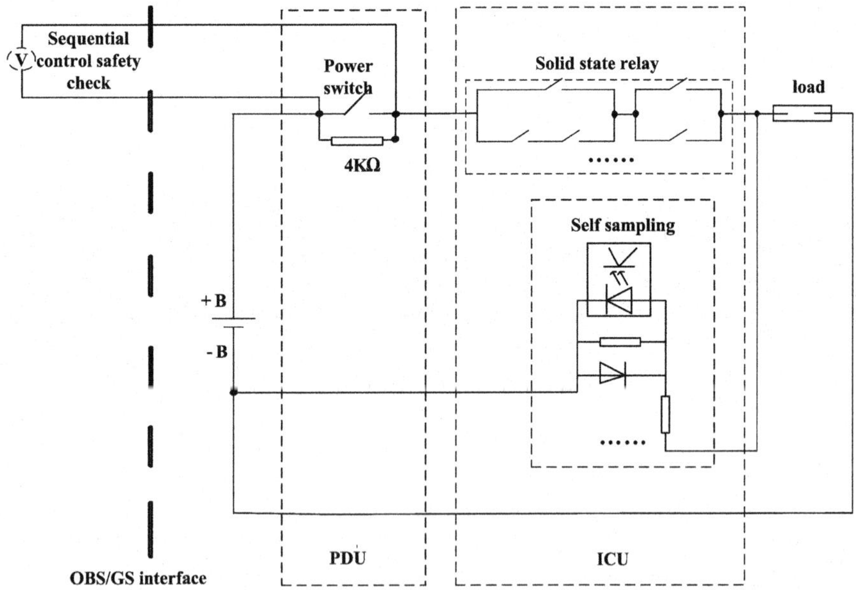

Fig. 4.14 Sequential control safety check for solid-state relay leakage current

assembly errors of the stages. If levelness difference exceeds the allowable value, measures are required to reduce the difference, and ensure that the launcher tail will not collide with the launch tower or launch pad due to the large initial attitude angle, particularly when a rear wing is installed on the launcher [9].

Perpendicularity adjustment is accomplished by two levelness measuring instruments mounted onboard. The upper end of the first-stage engine frame is selected as the launcher base plane, and the levelness measuring instrument is installed at the lower end of the frame, located in the crossbeam heading I, III, IV quadrant. The difference between the upper and lower end must be less than $1'$, selecting the least difference among the three low ends as the final instrument installation plane.

The upper plane of the IMU damping bracket is selected as the installation plane for another levelness measuring instrument. Launcher perpendicularity is adjusted depending on the measurement results, as shown in Fig. 4.15 (for simplicity, only one of the three first-stage instrument planes are shown).

When the launcher is tilted, the horizontal measuring instrument installed at the first stage will sense the inclination angle, α, and this is converted into a voltage, Us, applied to the levelness measuring instrument, which converts it to a digital signal, displaying the tilt direction (I or III, II or IV) and angle. The instrument also transmits Us to the launch pad automatic leveling system, which automatically adjusts the launch pad to correct the launcher perpendicularity.

The second levelness measuring instrument only measures the levelness where the IMU is installed, and that signal is not used for perpendicularity adjustment, but the levelness measurements are compensated for navigation.

4.1.3.4 Aiming Function Test

Launch vehicle navigation is based on the launch inertial coordinate system or launch coordinate system. The inertial navigation is fully automatic and recursive, and the initial states need to be determined first. These initial states or parameters include the initial position, velocity, and attitude in the coordinate system, which are determined by the pre-launch alignment operation with external support or known reliable information.

Most initial parameters can be determined in advance, such as the launch point geographical coordinates. Launch direction is measured or controlled by the aiming system before liftoff. For the inertial stabilization platform, the aiming system aligns with the prism installed on the body in the platform; when a deviation occurs, the body in the platform is rotated based on feedback to eliminate the deviation. Therefore, the X-axis in the platform coordinate remains aimed along the launch direction. However, the launcher does not rotate. Therefore, the Y_1 axis in the launcher body coordinate system does not align with the launch direction, and the frame angle of the platform will produce a roll attitude angle deviation. When the launcher lifts off, the platform body is stable in inertial space, and the platform coordinate system becomes the launch inertial coordinate system. Thus, the attitude

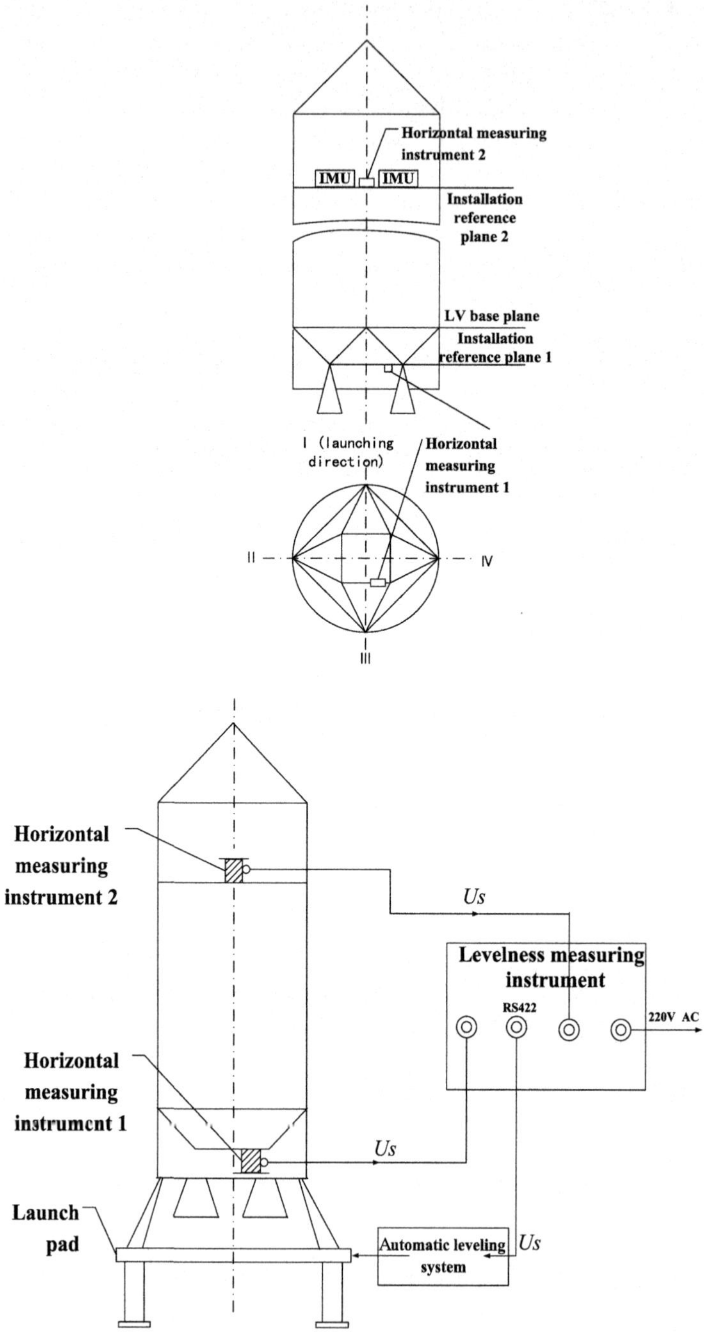

Fig. 4.15 Perpendicularity adjustment system block diagram

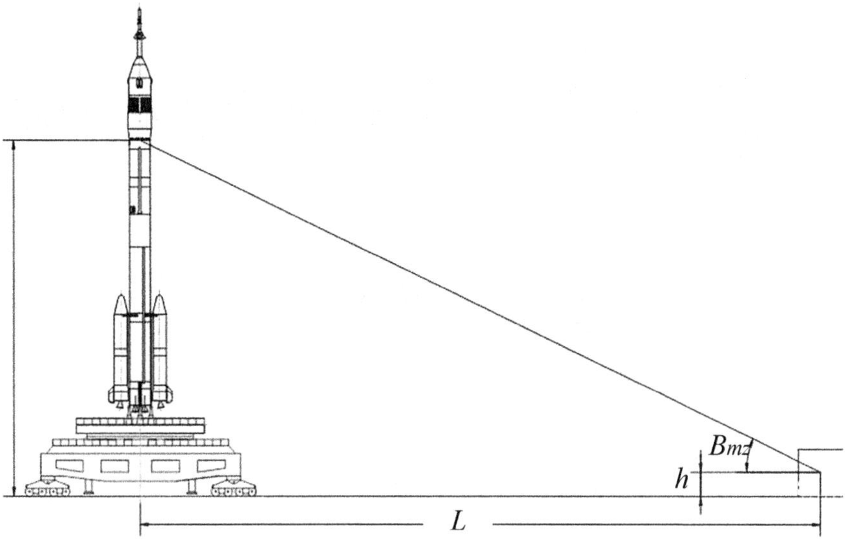

Fig. 4.16 Long-distance direct aiming scheme

control system rotates the launcher to eliminate the roll attitude angle deviation, aligning the Y_1 axis with the launch direction.

Since the inertial stabilization platform is gradually replaced by the IMU, IMU aiming is emphasized in this section. The aiming system comprises aiming equipment, launch pad, perpendicularity adjusting equipment, IMU with prism installed, software and hardware to calculate levelness and azimuth deviation, ground marking, supporting facilities, etc. Figure 4.16 shows a typical scheme with a fixed aiming room on the ground, where the aiming equipment is mounted in the opposite direction to launch at some distance L.

Figure 4.17 shows the pointing error on the aiming scaleplate, where point 9 is the aiming point. The reference geodetic azimuth aiming to point 9 is closest to the theoretical launch direction, A_0. Table 4.3 shows the intended workflow for the aiming scheme shown in Fig. 4.16.

Figure 4.18 shows an alternate close-range straight aiming scheme. The aiming device is mounted directly on the fixed service tower (e.g., the nth layer), at approximately the same height as the IMU.

Aiming is used to measure the deviation between the actual pointing direction of the IMU and the theoretical launch direction. This deviation will be eliminated by rolling control after takeoff and then the LV body will align with the launch direction. The actual azimuth of the IMU is calculated as follows:

$$A_l = A_b + \Delta A_a + (\beta_s + \psi_0)\tan B_{mz} + \alpha_s, \qquad (4.18)$$

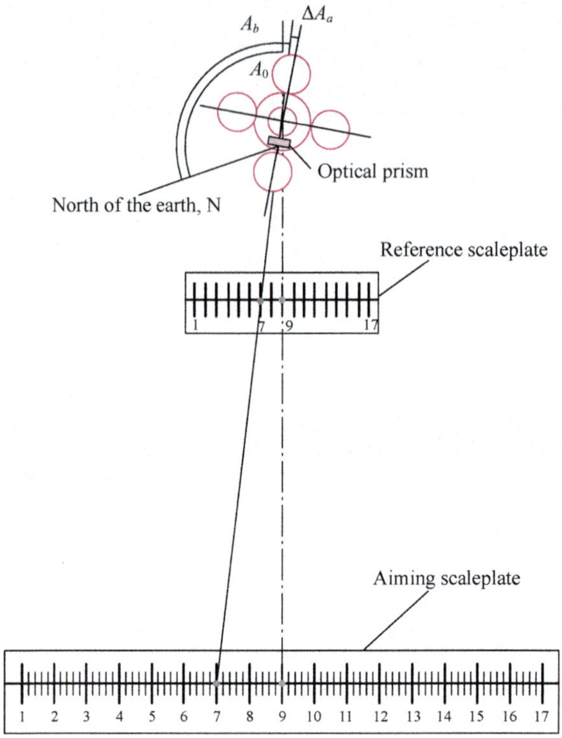

Fig. 4.17 Aiming and base scaleplate

Table 4.3 Typical workflow for long-range launch vehicle aiming

Step	Action	Details
1	Launcher is transferred to the launch center, and launch pad is roughly redirected	1. Adjust launcher perpendicularity 2. Align laser sight reference beam to the theoretical direction 3. Adjust laser sight elevation angle to ensure the reference beam points to the right angle reflecting prism on the IMU 4. Rotate the launcher body to ensure the laser beam aligns vertically with the prism Initial aiming is complete
2	Before fuel filling	1. Align laser sight reference beam to the theoretical launch direction 2. Adjust laser sight elevation angle to ensure the reference beam points to the prism on the IMU 3. If necessary, rotate the launcher body to ensure the laser beam aligns vertically with the right angle reflecting prism on the IMU 4. Control system calculates and records the IMU levelness and aiming deviation Pre-aiming is complete

(continued)

Table 4.3 (continued)

Step	Action	Details
3	Rocket is loaded before launch	1. Adjust launcher perpendicularity 2. Align laser sight reference beam to the theoretical launch direction 3. Adjust laser sight elevation angle to ensure the reference beam points to the prism on the IMU 4. Action the following operations as required based on the prism azimuth angle change before and after fuel filling a. If the change is small, move the guide rail of the laser sight to ensure the laser beam aligns with the right angle reflecting prism b. If the change is relatively large, choose another marked point on the aiming scaleplate and corresponding reference scaleplate, and reconfirm the reference azimuth Repeat step 4 as required to ensure the laser beam aligns vertically with the right angle reflecting prism c. If the change is very large, turn the launcher body to ensure the laser beam aligns vertically with the right angle reflecting prism 5. Record and transmit the reference azimuth $A_b, \Delta A_a$, and elevation angle B_{mz} to the control system 6. After the IMU is stable and aiming complete, transmit the aiming and the prism installation deviation to OBC by TLCS. Calculate the required roll program angle γ_0 by the OBC based on the deviation and IMU levelness 7. Two hours before launch, recalculate IMU levelness, verify $\Delta\gamma_0$, and determine whether the aiming elements should be reloaded depending on the current aiming situation 8. Thirty minutes before launch, operators check optical alignment status and evacuate. Alignment data is monitored by control center staff until liftoff

where A_l is the azimuth angle of the actual direction of the $-O_sY_s$ axis in the IMU, i.e., the actual geodetic azimuth at launch; A_b is the azimuth of an external reference used for the aiming device; ΔA_a is the difference between the azimuth of the prism normal and A_b, where the difference is measured by the aiming device; B_{mz} is the aiming elevation angle; α_s is the prism installation error relative to the direction of

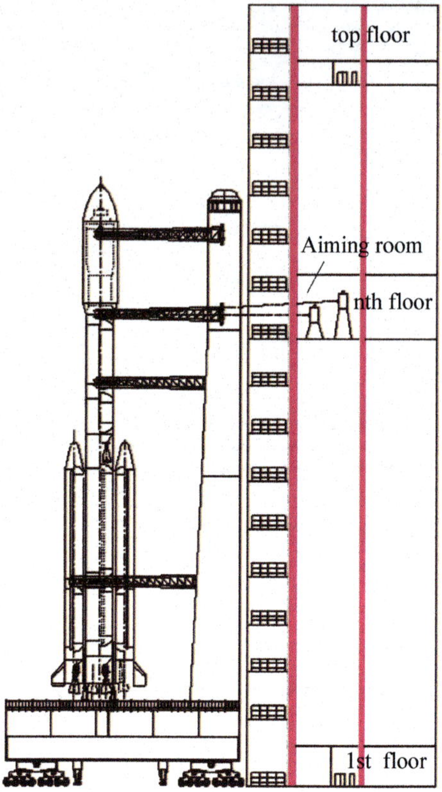

Fig. 4.18 Close-range horizontal aiming scheme

rotation around the O_sX_s axis in the IMU coordinate system, where positive is defined as counterclockwise when looking down; β_s is the prism installation error relative to the direction of rotation around the O_sY_s axis in the IMU coordinate system; and ψ_0 is the IMU levelness.

The roll angle is

$$\Delta\gamma_0 = A_l - A_0 \tag{4.19}$$

where $\Delta\gamma_0$ is the roll angle, which is reloaded to the OBC before launch, and $\Delta\gamma_0$ polarity is defined as positive when the right-hand thumb aligns with the direction of the OX_1 axis in the launcher body coordinate system and the roll direction is the same as the fingers when held tightly (the polarity in Fig. 4.19 is positive); and A_0 is the theoretical launch direction.

The launch inertial coordinate system established at takeoff based on IMU measurement information is a mathematical platform. Besides the pointing error, the levelness error of the IMU is also considered, as discussed in Eqs. (4.20) and (4.21). For physical platforms set by the inertial stabilization platform, the levelness and aiming errors are guaranteed by the platform itself.

Fig. 4.19 The rolling program angle polarity definition (looking down from the top of the launch point)

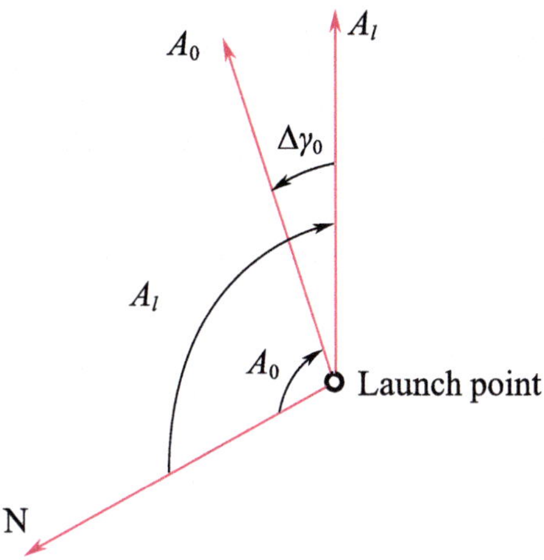

The IMU levelness error is mainly caused by installation error and structural deformation. When the launcher is stationary, levelness is measured by its own acceleration meters as

$$\Delta\varphi_0 = -\frac{\sum^{\Delta T} \delta\bar{w}_{y1}}{g_0 \cdot \Delta T}, \tag{4.20}$$

$$\psi_0 = \frac{\sum^{\Delta T} \delta\bar{w}_{z1}}{g_0 \cdot \Delta T}, \tag{4.21}$$

where $\sum^{\Delta T} \delta\bar{w}_{z1}$, and $\sum^{\Delta T} \delta\bar{w}_{y1}$, are the apparent velocity increments in ΔT seconds along the LV $z1$, and $y1$ axes, respectively; $\Delta\varphi_0$, and ψ_0 are the average levelness errors, respectively, over ΔT seconds in pitch and yaw direction.

In the case of wind before launch, which can shake the launcher slightly, levelness error can be identified more accurately by filtering accelerometer and gyro data.

4.2 System Level Tests

Every launcher requires specific appropriate test cases for system level tests. If the OBS is very complex, and system level checks were very difficult to accomplish using ground facilities. Therefore, an onboard test scheme was proposed, with all test data recorded and transmitted to the ground facilities for analysis [10]. In the test scheme of the space shuttle, flight software was extensively modified to suit test requirements, then verified in the Shuttle Avionics Integration Laboratory (SAIL), where some simulation models were applied. Only once the software passed the various checks was it allowed to participate in system level testing at the Kennedy Space Center.

However, the flight software cannot complete all the required tests of the space shuttle, so the other two software packages were developed: vehicle utility (VUSW) and dynamic integrated test (DIT). VUSW primarily connected the GS for generating data required for testing. DIT simulated dynamic in-flight sensor data, since these sensors only output static or fixed data during ground tests which are insufficient for system level testing. With these modifications, the space shuttle could simulate flight states on the ground. DIT inserted simulation data into the shuttle's bus, and stored the data on the tape. An important issue is to avoid instructions and codes that could cause safety issues or equipment damage during testing.

But there were still some differences remained between system level testing and real flight situations. Therefore, NASA proposed several improvement measures as follows.

1. The test code was embedded into the flight software, and some parameters were modified to ensure test codes or branches were called.
2. An independent test package, similar to VUSW, was designed for the tests.
3. Completely new software was designed to replace the flight software for testing.

However, the drawback of method 3 is that the flight software was not checked in ground tests. Methods 1 and 3 are also commonly considered in system level testing in China [11], and we introduce method 1 in subsequent sections.

4.2.1 General Check

All equipment are examined in the general check, which simulates the whole flight process for system level testing, and it is also referred to as the flight simulation test. This is the main verification method for the overall system in the assembly department and launch site. It is easy to simulate open-loop control processes, such as perturbation guidance, but testing closed-loop controls, such as iterative guidance, is somewhat more complex.

4.2.1.1 Perturbation Guidance Flight Simulation Test

Navigation Equations During Flight

The navigation equations during flight are as follows.

$$\begin{cases} g_\alpha = -\frac{fM}{r^3}(\alpha + R_{0\alpha}) - J \cdot \frac{fM}{r^2} \cdot \left(\frac{R_a}{r}\right)^2 \cdot \left[\left(1 - 5 \cdot \sin^2 \varphi_{dx}\right) \frac{\alpha + R_{0\alpha}}{r} + 2 \cdot \sin \varphi_{dx} \cdot c_\alpha\right] \\ V_\alpha = V_{\alpha,-1} + \Delta w_\alpha + \frac{1}{2}\left(g_\alpha + g_{\alpha,-1}\right) \cdot \tau, \end{cases} \tag{4.22}$$

where $\alpha(x, y, z)$ represents the position along the three axes in the launching inertial coordinate system,suffix 'α' represents the component of the corresponding physical parameter, g_α is the gravitational acceleration calculated according to gravitation, $R_{0\alpha}$ is the distance from the launch point to the Earth center, r is the distance from the LV centroid to the Earth center, R_a is the equatorial radius, ϕ_{dx} is the geocentric latitudes, Δw_α is the apparent acceleration increment measured by IMU, fM is the Earth gravitation constant, J represents the coefficient related to the Earth $J2$ coefficient, suffix '-1' represents the corresponding physical parameter of the previous cycle, τ is the sampling period, c_α is the coordinate transformation coefficient,

$$\begin{cases} c_x = \cos A_0 \cos B_0 \\ c_y = \sin B_0 \\ c_z = -\sin A_0 \cos B_0, \end{cases} \tag{4.23}$$

where A_0 is the launching direction, B_0 is launch point geodetic latitude.

The launcher's acceleration is the sum of the apparent and gravitational acceleration. When the IMU is resting on the surface of the Earth, the apparent acceleration it senses = 1 g, and gravitational acceleration is -1 g, hence acceleration = 0, neglecting any measurement error. Therefore, the cutoff equation cannot be met. To simulate the cutoff process for the ground test, gravitational acceleration was halved,

$$\begin{aligned} g_\alpha = & -\frac{fM}{2r^3}(\alpha + R_{0\alpha}) - J \cdot \frac{fM}{2r^2} \\ & \cdot \left(\frac{R_a}{r}\right)^2 \cdot \left[\left(1 - 5 \cdot \sin^2 \phi_{dx}\right) \frac{\alpha + R_{0\alpha}}{r} + 2 \cdot \sin \phi_{dx} \cdot c_\alpha\right] \end{aligned} \tag{4.24}$$

which is equivalent to generating 0.5 g acceleration in the vertical axis of the launcher coordinate system, and the IMU can also sense the Earth's rotation. Then the perturbation cutoff threshold at the theoretical shutdown time can be calculated based on the navigation and cutoff equations sensing the 0.5 g acceleration and the Earth's rotation, and uploaded to the OBC for testing, ensuring the cutoff command will be sent at the approximately theoretical time during ground testing.

For example, suppose the first-stage engines shut down at time T_1, depending on the flight range; second stage main engine shuts down at T_2, depending on launcher velocity; and swing engines shut down at T_u, depending on the semi-major axis. Range, velocity, and semi-major axis values at these time points can be determined for the half-gravity case in advance, and provide the thresholds for the cutoff control that are expected to occur during the test, in principle, at the same T_1, T_2, and T_u. The flight software process is shown in Fig. 4.20.

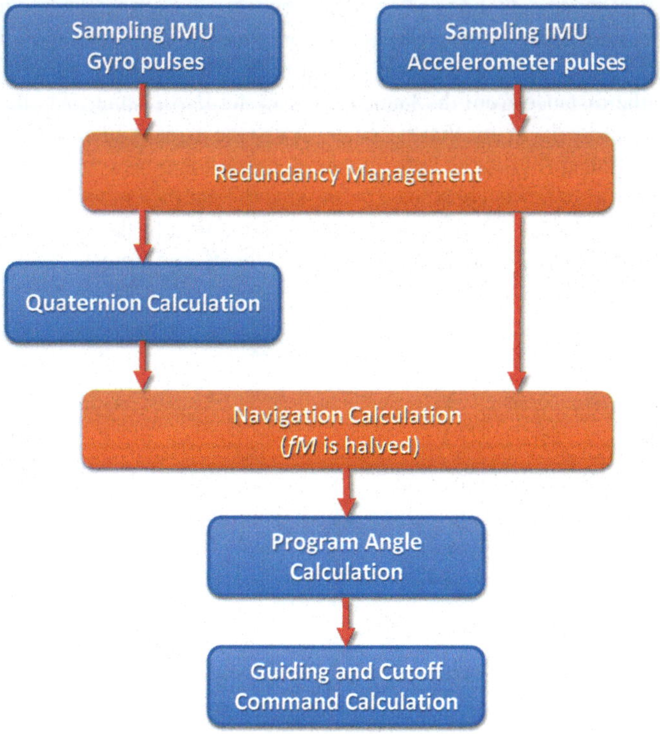

Fig. 4.20 Flight simulation process for perturbation guidance method

Test Signals Considered

The following signals were considered for ground testing.

1. Program angle.
   Since there is no requirement to simulate the actual flight program angle during the test, the angle can be set artificially, generally to meet polarity check requirements. The aim is to check the whole link from the program angle to the servo mechanism feedback. The program angle used in the check can be referenced to the attitude control system test in Sect. 4.1.1.1.
2. Guiding control.
   Guiding control ensures the launcher follows the desired flight trajectory. It can also be simulated artificially during testing, to check the polarity from the guiding control to the servo feedback. They are applied independently at times to distinguish them from the program angle.

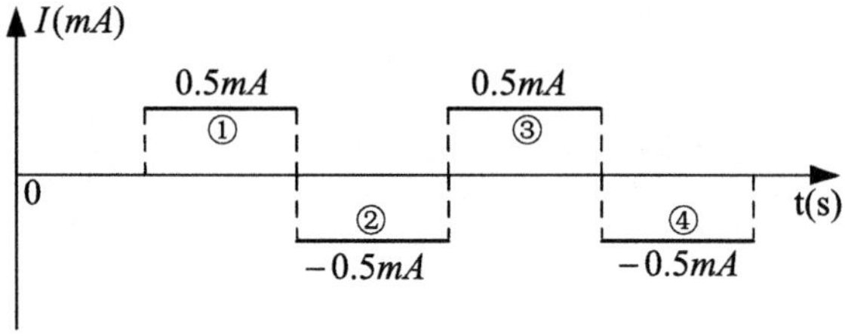

Fig. 4.21 Typical torque command

3. Torque command.
   Torque can be applied to the accelerometers (or gyros) to generate differences between the redundant meters, and hence trigger and check OBS FDI function. Figure 4.21 shows an accelerometer torque test example for TMR design, where

   ① is the positive torque stimulation for AY, BY, and CZ accelerometers;
   ② is the negative torque stimulation for AY, BY, and CZ accelerometers;
   ③ is the positive torque stimulation for CY, AZ, and BZ accelerometers; and
   ④ is the negative torque stimulation for CY, AZ, and BZ accelerometers.

   where A, B, and C are IMU indicators, and *Y*, *Z* are axis indicators. For the first two cases, two *Y*-axis accelerators (AY, BY) have the same output with the stimulated torque, so the third (CY) is judged as faulty due to inconsistency with the other two. Only the CZ accelerometer is torque stimulated, hence it has different output from the other *Z*-axis accelerators (AZ, BZ), so is identified as faulty.
   The situations are similar for the latter two cases.
4. Rotating platform instructions.
   The IMU is mounted on a rotating platform in the laboratory and assembly factory, and the attitude angle can be simulated by turning the platform, with corresponding accelerometer responses. Thus, the attitude to servo feedback polarity and static gain can both be checked. Shutdown control during system level testing is timed in this situation.

The above test cases are appropriate for perturbation guidance, but under iterative guidance, the flight program angle is calculated in real-time, and there is no predefined desired flight trajectory, so guiding control is not required.

4.2.1.2 Iterative Guidance Flight Simulation Test

The launcher is stationary during the testing, hence the program angle does not affect LV attitude, and since the IMU is also still on the ground, its output will not contain the effects of guidance (program angle) and attitude control command. Thus, the closed-loop control is broken. Few studies have considered a closed-loop general check scheme [12, 13], because the requirement is relatively light for perturbation guidance, but it must be considered for closed-loop or iterative guidance.

Since there is no thrust during the general check, and the IMU data does not respond to guidance commands, then no matter how the control system operates, the launcher state remains fixed, and the cutoff equation is never met. Although IMU output can also be stimulated through the torquer (not suitable for optical gyros), it is difficult to maintain synchronization between the output and guidance commands.

Therefore, the following two schemes are mainly adopted for the closed-loop general check.

Data Replacement

The method is similar to the space shuttle DIT test [10]. A trajectory is selected for the ground simulation, such as a standard trajectory with no disturbance or with some specific disturbances.

The program angles, or guidance commands, are implemented by the attitude control system to drive the servos swinging engines. In the simulation environment, the swing changes the launcher's motion state, which is sensed by the IMU, and the new IMU output can be calculated and provides input for the guidance method in the next control cycle. Thus, a closed-loop simulation is formed, and IMU measurements and program angles are recorded for each control cycle, where the IMU measurements are considered to be virtual inertial signals and the program angles to be the benchmark.

The virtual inertial signals are stored in the flight software during the general check, replacing sampled IMU data when iterative guidance starts. Guidance commands calculated by the flight software based on these virtual inertial signals can then be compared with the benchmark, i.e., the program angles corresponding to the virtual inertial signals recorded in the simulation. Figure 4.22 shows the operational principle.

Thus, the flight software iterative guidance method can be examined. For example, if the guidance control laws were incorrectly coded in the flight software, the program angles calculated from the virtual inertial signals will differ from the benchmarks, and cutoff conditions may also not be met.

However, this approach also has several problems. Real IMU outputs are replaced with virtual inertial signals, and the IMU is no longer involved in the calculation, or system level testing. In addition, the virtual inertial signals stored in the OBC occupy a large amount of memory (the US space shuttle adopted data insertion through the data bus to avoid influencing OBC and flight software).

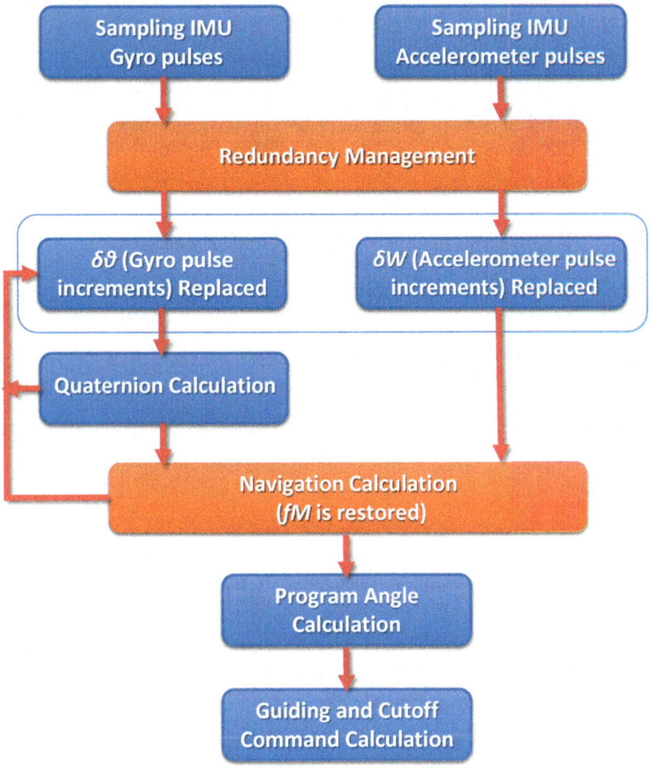

Fig. 4.22 Flight simulation process for iterative guidance method with data replacement

Data Superposition

Figure 4.23 shows the data superposition principle [13]. Ground terminal simulation results and IMU outputs are both sent to the OBC. The ground terminal samples the servo outputs, calculates the launcher state of the next cycle, and converts launcher acceleration and attitude to what the IMU would sense, w_i, i.e., the theoretical IMU output.

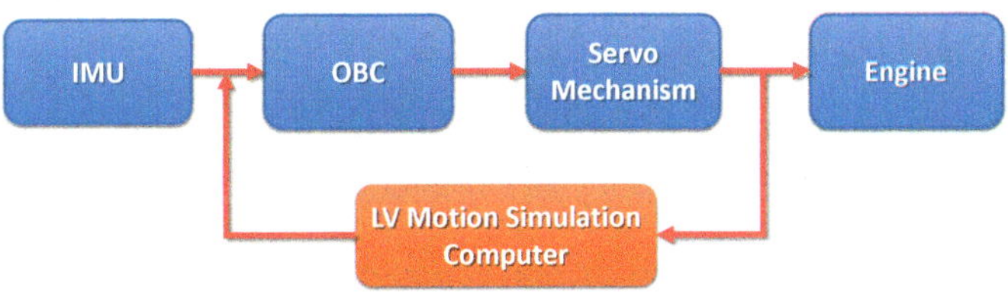

Fig. 4.23 The flight simulation principle for iterative guidance method with data superposition

Suppose the actual IMU output during the test is w, sensing local gravity and Earth's angular rotation. The IMU output consists of two parts: w_s, which can be determined from IMU calibration parameters; and δ, which is determined by meter random errors, and only the limits can be judged. Therefore, $w = w_s + \delta$, and the ground terminal transmits $(w_i - w_s)$ to the OBC. The total signal sampled by the OBC, w_{sample}, is

$$w_{\text{sample}} = w_i - w_s + w = w_i - w_s + w_s + \delta = w_i + \delta. \tag{4.25}$$

If the OBC only samples w_i, the program angle calculated by OBC flight software should be consistent with that by ground terminal simulation. However, when δ is introduced, the results are no longer identical. The difference can not only verify algorithm accuracy but also IMU accuracy. This is discussed further in the closed-loop comprehensive test in Sect. 7.4.

Flight Simulation States

Based on extensive practical experience and engineering principles, the control system has three general check states during flight simulation, as follows.

1. Power transfer state (or 'state I' for short).
   OBS and GS connectors are detached following the planned time sequence, and power transfer control is checked. Thus, powering is transferred from ground power sources to onboard batteries. Sometimes, the batteries are substituted by direct connection to a ground power supply.
2. Non-power transfer state (or 'state II' for short).
   The OBS is powered by a ground power supply during testing with no power transfer control, and all connectors between OBS and GS remain coupled.
3. Emergency shutdown state (or 'state III' for short).
   The pre-launch process is strictly simulated. At some time after the ignition command is sent, the TLCS automatically transmits the emergency shutdown control command. This command reaches the OBS through the connectors between the OBS and GS, and the engines will be cut off.
   This is used to simulate an emergency state during liftoff. Normally, the command could not reach the OBS due to the connectors between the OBS and GS being decoupled after LV liftoff. If the connectors remain coupled because the launcher cannot reach a specific height (regarded as no-takeoff, usually caused by insufficient thrust or ignition failure), the OBS will receive the shutdown command and trigger an emergency process.

The different states have different objectives. During debugging, or when the system has not been sufficiently checked, state II is preferable. Retaining all GS and OBS connectors provides sufficient measurement and monitoring points, the OBS can be comprehensively examined, and power-off control is more convenient under emergency conditions. However, with the wide adoption of BIT, most test data are relayed to the GS in digital form through communication lines. The requirement for

Table 4.4 The engine shutdown mode in-flight simulation test

		Booster	First core stage	Second core stage	
				First cutoff	Second cutoff
State I (power transfer state)		Backup cutoff based on first-stage shutdown command	Guidance cutoff	Guidance cutoff	Guidance cutoff
State II (non-power transfer state)	A	Exhaustion (the first is booster 1 and 3, then booster 2 and 4)	Guidance cutoff	Guidance cutoff	Guidance cutoff
	B	Backup cutoff based on 1st stage shutdown command	Guidance cutoff	Guidance cutoff	Exhaustion
	C	Exhaustion (the first is booster 2 and 4, then booster 1 and 3)	Timing cutoff	Timing cutoff	
State III (emergency shutdown state)		Emergency shutdown		–	–

direct sampling by GS facilities has significantly declined; even test connectors are rare. Thus, this state no longer provides any advantages and may be omitted.

For tests including power transfer control, since all OBS and GS connectors are separated, the OBS cannot power-off itself. If the OBS is powered on by batteries, a temporary power control cable could be connected to the onboard power distribution unit (PDU) directly from the GS, and the switch off command can be sent via this connection to PDU. If this temporary connection is not possible, at least one GS and OBS connector must remain coupled for power-off control. If the OBS is not powered by batteries, i.e., through a direct connection to ground power supplies via an independent cable, power-off after power transfer control can be achieved by switching off the ground power supply.

The emergency shutdown control is used to check the whole launch process. It starts from the launch preparation phase in strict accordance with the actual process and time expense, and the TLCS operates in the launch state (the other two tests are implemented in the check state).

Different test cases are designed into the general check, to examine various cutoff modes, as shown in Table4.4.

4.2.2 Engine and Propellant System Tests

This is a special ground test, involving many subsystems, but every subsystem only partially participates.

4.2.2.1 Engine Test Classification

For newly developed engines, engine tests occur many times before the LV inaugural flight. The control system participates in these tests to control pressure, and engine ignition and cutoff. The process is similar to the general check, where equivalent devices (loads) are replaced by real valves and pyrotechnics, and not all control system equipment is required. This section provides a brief introduction.

Engine tests are usually planned on three levels, as follows.

1. Engine runs independently and separately.
   All propellant filling, pressurization control, ignition, shutdown control, etc., are independently performed by the engine manufacturer.
2. Engine and control system integration test.
   Since engine ignition, shutdown, and swing are completed in-flight by the control system, it is essential to verify the integration between these subsystems. However, the engine manufacturer is still responsible for propellant filling and pressurization control.
3. Whole system test.
   This is performed in a special test bed. Pressure control, propellant delivery, TLCS, and telemetry systems are all involved in addition to the engine and control systems. The propellant filling is completed as before, but pressure control is managed by the OBS. The process is the same as that before launch, all activities related to propellant treatment before ignition are managed by GS, and then replaced by the OBS when the engines run. The telemetry system provides monitoring during the test.

4.2.2.2 Test Facility Composition

Figure 4.24 shows a typical control system and TLCS configuration for first-stage engine test. The cabin of the first stage contains all equipment for engine control. Batteries and PDUs are responsible for powering servo valves, pressure sensors, and onboard devices. Pressure sensors measure pressure within the propellant tanks, and exhaust sensors transmit shutdown requests. Valve excitation commands are sent by the ICU for engine ignition, cutoff, and pressure control. The servo controller drives servo mechanisms to swing the engine.

The required equipment installed in the instrument cabin during flight are put on the test bed, because this cabin is not involved in engine testing. For example, the OBC is configured to control engine swing through SCUs.

Launch control is managed by the GS, which is divided into front-end and back-end equipment, but only those required for engine control are included in the test. Front-end equipment includes ground power supplies (AC and DC), PLCs, signal interconnection assemblies related to launch, and the communication computer between OBS and GS. Back-end equipment includes desktop computers for virtual display, virtual launch panels, and process control facilities.

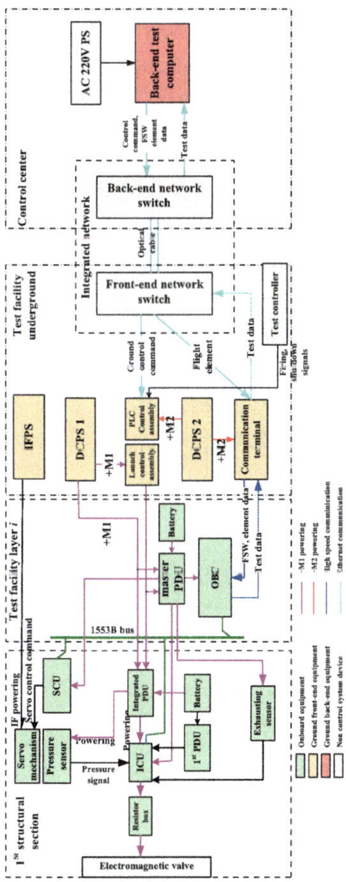

Fig. 4.24 Control system block diagram for engine test

4.2.2.3 Test Process

Figure 4.25 shows the test process. To ensure smooth progress, each valve is preferably tested individually to check on–off function, then examined by coherent sequential commands. After all tests are successfully completed, the engine runs from ignition to shutdown (sometimes emergency shutdown is also evaluated), and power-off all system. Shutdown can be in normal or emergency mode.

4.2.2.4 Test Information Flow

Figure 4.26 shows the test information flow. Flight software and pressure control elements are uploaded to the OBC through a ground communication terminal and LV–TLCS interfaces. The OBC transmits the pressure control elements to other ICUs depending on the process control commands. ICUs complete the pressure control based on real-time sensor data, and receive sequential commands from the OBC to drive the valve to start or stop the engine. The OBC also transmits attitude control commands to SCUs, which drive the servo mechanism to swing the engine. Engine swing angle, sequential command timing, and pressurization valve on/off times are self-checked, collected by the ICUs and OBC, and sent to the GS. This scheme greatly simplifies on-site test facilities.

A PLC controls the PDU for power-on, power switch, and power-off, and OBS BIT information and test states are displayed on the back-end terminals.

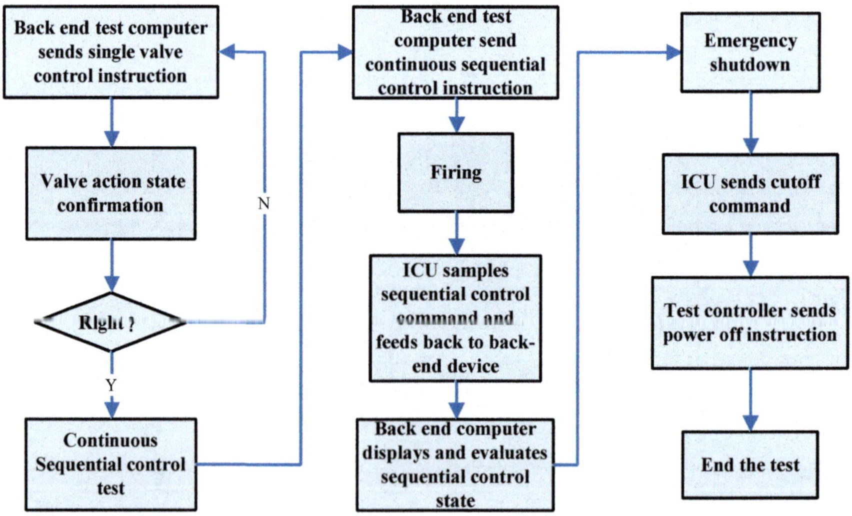

Fig. 4.25 Engine system test process

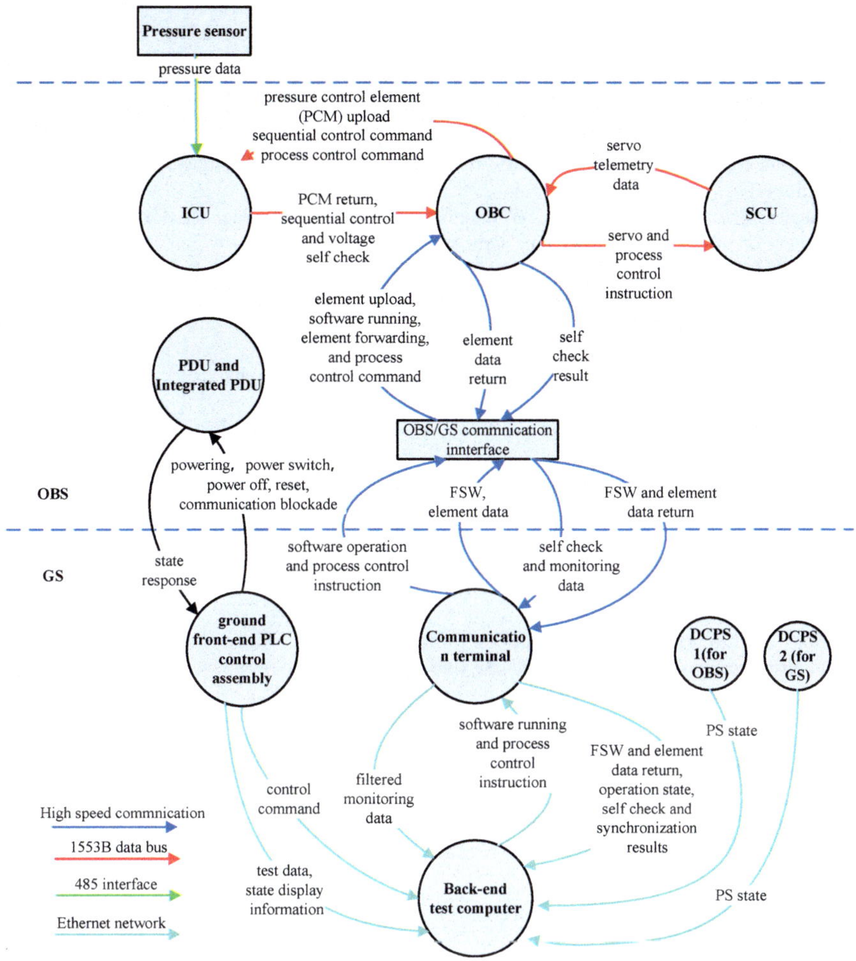

Fig. 4.26 Engine system test information flow

Engine tests for the different stages are similar. When only the control system is involved, ground test facilities can be further simplified, e.g., back-end computers can control onboard equipment directly through a network and the front-end facilities are omitted.

References

1. Song ZY (2014) High reliability design for launch vehicle control system. China Aerospace Publishing House, Beijing, pp 262–297 (in Chinese)
2. Xu YW (editor in chief) (1992) Control system (3). Missile and aerospace collection—liquid ballistic missile and launch vehicle series. China Aerospace Publishing House, Beijing (in Chinese)
3. Fan JS, Zhang HX, Yang JH (2007) Automatic test and fault diagnosis for subsystem test system of missile's control system. Ordnance Ind Autom (10):72–74 (in Chinese)
4. Wang JH (1999) The design of LM-3A launch vehicle dynamically measuring-control system. Missiles Space Veh (4):23–27 (in Chinese)
5. Sun NS (2003) Application of redundant design technology in the flight control system of launch vehicle (2). Aerosp Control 21(2):68–80 (in Chinese)
6. Zeng GS, Shen WG, Shi L et al (2005) High reliability triple redundancy servomechanism system. Aerosp Control 23(1):35–40 (in Chinese)
7. Zhao SJ, Zhao YX, Jiang QY et al (2012) The 2nd stage servo-mechanism system for liquid oxygen-kerosene manned launch vehicle. Manned Spaceflight 18(5):5–11 (in Chinese)
8. Liu XY, Peng QS (2009) Solution and realization of rocket-borne pyrotechnics auto-test. J Projectiles, Rockets, Missiles Guidance 29(5):232–234 (in Chinese)
9. Wu MQ, Liu HB, Yang XR et al (2013) Method and development of launch vehicle launch pad perpendicularity adjustment. Missiles and Space Veh (2):30–35 (in Chinese)
10. Mayer J The space shuttle vehicle checkout involving flight avionics software. AIAA, AIAA81-2141
11. Song ZY (2014) High reliability design for launch vehicle control system. China Aerospace Publishing House, Beijing, pp 236–237 (in Chinese)
12. Zhang J, Sun NS (1999) Exciting methods of inertial instruments in closed-loop flight simulation of launch vehicle. Aerosp Control (4):27–30, 37 (in Chinese)
13. Zhou H, Song ZY (2013) The design and research of closed loop general testing technique based on BMU and LVDS bus. Aerosp Control 31(3):45–49 (in Chinese)

Chapter 5
Simulation Tests

This section introduces simulation tests used for the control system. Simulations are based on control theory, launcher physical and mathematical models, and advanced computer technology, and provide a comprehensive method to verify the control system design, particularly for the guidance algorithm, control laws, and related software and hardware [1–4].

Simulation tests are usually divided into three categories, as follows.

1. Model-in-the-loop (MIL) simulation, where the sensors, controllers, and actuators are modeled mathematically, and the simulation is performed on a calculation platform.
2. Software-in-the-loop (SWIL) simulation, where the control algorithm is coded and compiled, but not necessarily running on the target unit.
3. Hardware-in-the-loop (HWIL), also known as semi-physical simulation, where the software runs on the target unit, and real launcher equipment, such as controllers, actuators, etc., and some simulators or models, are included. For example, simulators provide IMU outputs, particularly accelerometers, and models represent engine load characteristics. Fault simulators or failure injection devices are also used to check redundant design.

Simulation tests have had important roles in design verification and validation, pre-launch technical support, operational troubleshooting, etc., [5, 6]. For example, NASA's Jet Propulsion Laboratory (JPL) developed the Integrated Test Laboratory (ITL) for the Cassini project. This is a HWIL simulation laboratory, where the command and data subsystem, attitude control subsystem, etc., are integrated in the forms of models or prototypes. The ITL simulation system is mainly used before launch to verify the system architecture, flight software, GSE; and during operation to check operational sequences, investigate abnormal conditions, train, etc. Another example is the Ares SIL that provides a real-time HWIL simulation test environment.

This chapter first reviews simulation technology in the aerospace industry, with particular attention to LVs, and then discusses simulation testing for LV control

Z. Song, *The Test and Launch Control Technology for Launch Vehicles*,
https://doi.org/10.1007/978-981-10-8712-7_5

systems. Basic simulation testing principles are described, and then the test system composition is presented to enable the reader to quickly understand the general concepts. Subsequently, we briefly introduce the model and boundary conditions commonly applied for LV simulation analysis, as this is the most common practice.

5.1 Brief Introduction to Aerospace Simulation

Simulations are divided into three categories, as discussed above, depending on the product status.

Model-in-the-loop simulation is widely used in LV design and verification. For example, NASA designed a simulation system for Ares rocket models [7]. Many different models are developed, covering such aspects as structural dynamic, quality characteristics, structural characteristics, nozzle dynamics, sloshing dynamics, atmospherics, centralized and distributed aerodynamics, gravity, booster separation motor, engines, reaction control systems, etc., and combinations of various relevant electronic equipment, such as flight computer, boost control power distribution unit, etc. These models were integrated to construct the virtual flight simulation.

MIL simulations are not only used to simulate flight processes, but are also widely used for operation and process simulations with deeper combinations of virtual reality [8]. The simulations can identify in advance if the operation is not appropriate, or has critical interferences, and maintenance issues.

MIL simulation used for algorithm verification is also known as mathematical simulation in China, and operation-related simulations have also gained significantly more attention recently. Figure 5.1 shows an example of the digital joint operation simulation for the new-generation Chinese LVs [9].

Software-in-the-loop (SWIL) simulation integrates various control algorithms into software products. However, SWIL and HWIL are relatively similar approaches. If the software is intended to be run on target units to be checked, it is also known as processor-in-the-loop testing for NASA. If SWIL simulations involve not only the processor or the OBC, but also other avionics, it is also known as HWIL simulation, no matter what the hardware scope is.

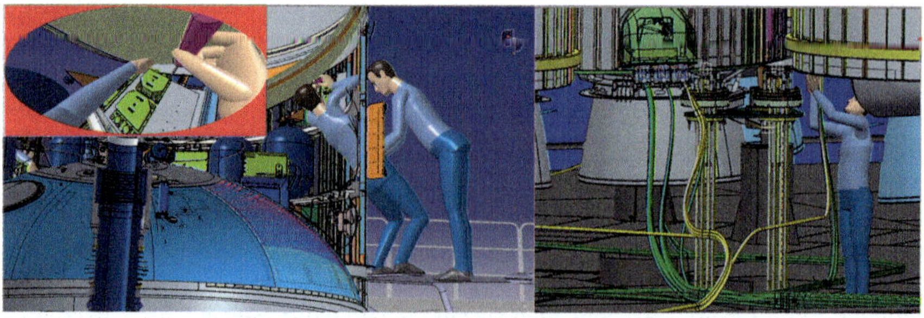

Fig. 5.1 Chinese new-generation launch vehicle digital joint operation simulation

Flight software tests in China are often performed on simulation testing platforms, comprising the OBC, where the software runs; other integrated simulators to produce various input signals; and various loads for the software outputs.

Hardware-in-the-loop simulation does not involve all hardware [10]. For example, it is difficult to create flight overload under ground test conditions, so the acceleration is usually generated by a simulator.

Simulations should use real boundary conditions as much as possible because this is a key input. Although MIL simulation is quite mature, many designs or manufacture deviations are difficult to model. Therefore, physical objects are widely used in large-scale simulation systems, particularly those related to the structure and mechanism. For example, the NASA Johnson Space Center constructed a 6-DOF dynamic test system to provide a closed-loop test environment for spacecraft rendezvous and docking [11]; and the Space Operation Simulation Center can verify aircrafts' relative navigation algorithms, spacecraft guidance and control algorithms, sensor performance, and structural and mechanical characteristics related to rendezvous and docking.

Chapter 4 preface discussed how system level tests are performed as two separate aspects in the Chinese aerospace industry: avionics integration tests and semi-physical simulation tests, which complement each other. Simulation scope is very broad, hence this section focuses on LV control system simulations. Chapter 7.4 presents a new system-in-the-loop (SysIL) test that combines electrical system integration tests and control system simulation advantages.

5.2 Basic Principles of the Rocket Control System Simulation Test

Mathematical simulation of LV guidance systems is very efficient, fast, and relatively easily implemented, where the body and control models are integrated and operated on the simulation computer. However, there are significant differences to the actual system, particularly between the simulation computer and target OBC. Therefore, a semi-physical simulation scheme is adopted, where the FSW runs in the OBC (target unit), OBC input and output signals are simulated by equivalent devices, and LV motion is calculated in the simulation computer. Guidance and attitude controls are closely coupled by applying closed-loop guidance. Therefore, 6-DOF simulation tests are preferable, to assess iterative guidance effects on flight stability, and the influence of the attitude control accuracy on guidance precision [12]. Figure 5.2 shows the 6-DOF simulation principle.

The main advantages of 6 DOF over classical 3-DOF simulations are as follows [12, 13].

1. Linearized small deviation increments equations are replaced by time-varying nonlinear full quantity equations.

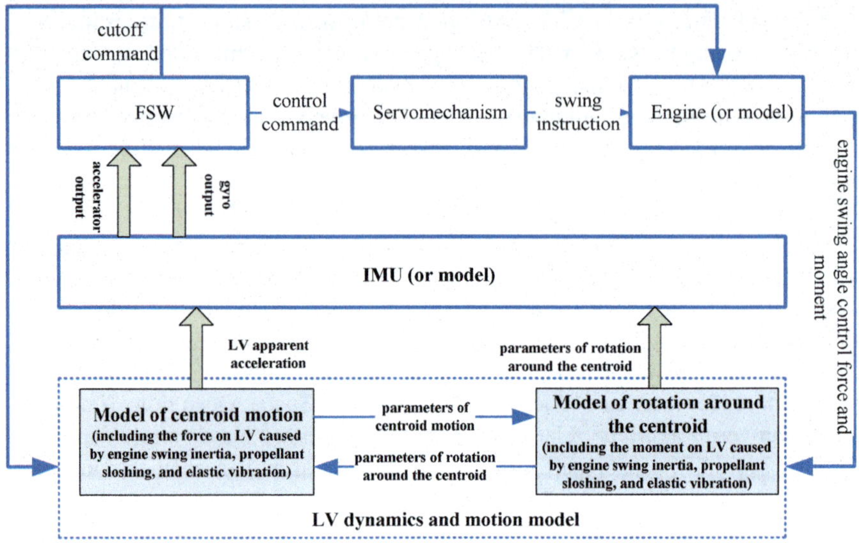

Fig. 5.2 6-DOF simulation block diagram

2. Coefficients of the body motion equation do not rely on the theoretical standard trajectory, and are derived from real-time calculations.
3. Guiding signals are not derived from theoretical values, usually the standard trajectory, but are replaced by the program angles calculated from the real-time guidance algorithm.
4. In addition to engine thrust and aerodynamic force in 3-DOF simulations, 6-DOF simulations can also consider engine swing, propellant sloshing, and elastic vibrations, all of which affect accelerometer measurements.
5. Body deformation influences on attitude measurement sensors may also be considered.

Motion of and around the body center is calculated on the simulation computer, as well as elastic and sloshing motions, structural disturbances, and engine thrust ripples. Environmental aberrations can also be simulated, such as steady, shear, or intermittent wind, etc. Thus, simulations provide more comprehensive evaluation of guidance and attitude control subsystem performances, and their coupled effects.

Real equipment or prototypes are often introduced due to various nonlinear characteristics that are difficult to model, such as OBC, IMU, servos, engines, their load, etc., whereas the LV body is modeled mathematically, producing a typical semi-physical simulation system as shown in Fig. 5.3.

The IMU is installed on the 3-axis-rotating platform, while the rate gyro is mounted on a single-axis-rotating platform. Depending on the body attitude, calculated by the simulation computer based on the LV's motion equation, the simulator controls the 3-axis-rotating platform to simulate the attitude, and the

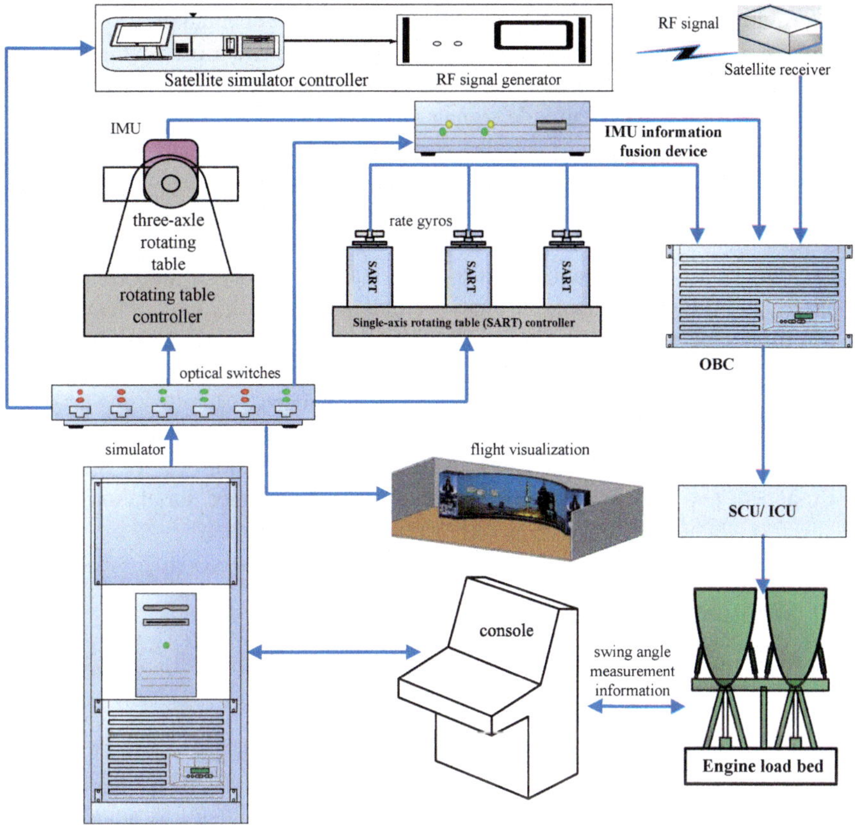

Fig. 5.3 An example of 6-DOF simulation experiment system

single-axis-rotating platform to simulate angular rate. Simulated launcher acceleration is also sent to the IMU information fusion device, replacing the output of IMU accelerometers. The fused acceleration, attitude, rate gyro outputs are sent to the OBC. Flight control software receives integrated navigation information, and completes redundant processing, navigation calculations, and guidance and attitude control calculations, then transmits commands to the SCU to swing the engines, and swing angles are measured and fed back to the simulator to determine LV current state, such as position, velocity, and attitude. Thus, a closed-loop simulation test system is created.

The engines are usually prototype devices, or simulators with equivalent moments of inertia and center of mass to the real engines, and installed on the load bed. The engine load bed is used to simulate nozzle forces and accurately simulate engine swing characteristics.

Optical fiber network-based memory sharing is adopted to ensure synchronization between multiple simulation devices, and various control commands can be transmitted in real time through the optical fiber. For example, LV motion parameters calculated by the simulator are mapped in real time to the rotating platform controller memory, and the controller rotates the platform to enable the IMU to output the desired attitude signals.

Redundancy management may also be tested in the simulation, where the key aspect is the IMU FDIR. Since FDIR requires a certain period, in the transitional, i.e., from when a failure is detected to being completely isolated; or development, i.e., when the output of the abnormal equipment does not exceed the threshold, phases, the faulty device data may also participate in navigation and control calculations, and this effect should be analyzed in detailed. Thus, inertial meter failure modes cannot be simply regarded as 0 (no output), constant value, or maximum value, and the time characteristic should also be considered, i.e., the fault development and propagation process should also be simulated.

Adaptability to thrust reduction, algorithm convergence, the probability of propellant exhausted, and flight stability due to the reduced control force, should all be examined, and this is a primary focus for iterative guidance simulation.

5.3 Simulation Modeling

5.3.1 Control System Equipment Models

Launcher control systems usually include IMUs, OBC, ICUs (drivers), SCUs, servos, power supplies (batteries), and PDUs. Some of these may be mathematically modeled and/or converted to a simulation model. If it is difficult to model, but a prototype is feasible, a combined model and prototype, or semi-physical simulation, is also acceptable.

5.3.1.1 Inertial Measurement Equipment

Inertial measurement equipment includes the IMU, rate gyros, accelerometer assemblies, etc. We specifically discuss IMU modeling, but approaches for other inertial equipment are similar. The IMU comprises gyroscopes and accelerometers, and we consider the laser IMU (LIMU) to illustrate the mathematical modeling.

A general mathematical model for laser gyro measurement can be expressed as

$$\begin{cases} (\Delta N_{gx+} \cdot K_{gx+} - \Delta N_{gx-} \cdot K_{gx-}) = D_{0x} + \omega_x + D_{xy} \cdot \omega_y + D_{xz} \cdot \omega_z \\ (\Delta N_{gy+} \cdot K_{gy+} - \Delta N_{gy-} \cdot K_{gy-}) = D_{0y} + D_{yx} \cdot \omega_x + \omega_y + D_{yz} \cdot \omega_z \\ (\Delta N_{gz+} \cdot K_{gz+} - \Delta N_{gz-} \cdot K_{gz-}) = D_{0z} + D_{zx} \cdot \omega_x + D_{zy} \cdot \omega_y + \omega_z, \end{cases} \tag{5.1}$$

where ω_x, ω_y, and ω_z are the angular rates (°/h) along the IMU three measurement axes, respectively. ΔN_{gx+}, ΔN_{gx-}, ΔN_{gy+}, ΔN_{gy-}, ΔN_{gz+}, and ΔN_{gz-} are the pulse numbers per unit time (P/s) of the three gyroscopes positive and negative output channels, respectively. K_{gx+}, K_{gx-}, K_{gy+}, K_{gy-}, K_{gz+}, and K_{gz-} are the gyroscopes pulse equivalency ($''/P$) for each channel, respectively. D_{0x}, D_{0y}, and D_{0z} are the gyroscopes zero-term drift coefficients (°/h), respectively. D_{xy}, D_{xz}, D_{yx}, D_{yz}, D_{zx}, and D_{zy} are the gyroscopes installation error coefficients, respectively.

Similarly, a general mathematical model for accelerometer measurement is

$$\begin{cases} \left(\frac{\Delta N_{ax+}}{K_{ax+}} - \frac{\Delta N_{ax-}}{K_{ax-}}\right) = K_{0x} + \dot{W}_x + E_{xy}\dot{W}_y + E_{xz}\dot{W}_z + K_{2x}\dot{W}_x^2 \\ \left(\frac{\Delta N_{ay+}}{K_{ay+}} - \frac{\Delta N_{ay-}}{K_{ay-}}\right) = K_{0y} + E_{yx}\dot{W}_x + \dot{W}_y + E_{yz}\dot{W}_z + K_{2y}\dot{W}_y^2 \\ \left(\frac{\Delta N_{az+}}{K_{az+}} - \frac{\Delta N_{az-}}{K_{az-}}\right) = K_{0z} + E_{zx}\dot{W}_x + E_{zy}\dot{W}_y + \dot{W}_z + K_{2z}\dot{W}_z^2, \end{cases} \tag{5.2}$$

where $\dot{W}_x$, $\dot{W}_y$, and $\dot{W}_z$ are the apparent accelerations (g_0) along the IMU three measurement axes, respectively. ΔN_{ax+}, ΔN_{ax-}, ΔN_{ay+}, ΔN_{ay-}, ΔN_{az+}, and ΔN_{az-} are the pulse numbers per unit time (P/s) of the three accelerometers positive and negative output channels, respectively. K_{ax+}, K_{ax-}, K_{ay+}, K_{ay-}, K_{az+}, and K_{az-} are the accelerometers pulse equivalency (P/($g_0 \cdot$ s)) for each channel, respectively. K_{0x}, K_{0y}, and K_{0z} are the accelerometers zero-term drift coefficients (g_0), respectively. E_{xy}, E_{xz}, E_{yx}, E_{yz}, E_{zx}, and E_{zy} are the accelerometers installation error coefficients, respectively. K_{2x}, K_{2y}, and K_{2z} are the accelerometers second term coefficients (g_0/g_0^2), respectively.

Gyro dynamic characteristics influence LIMU control system stability, which is generally described by a second-order differential equation,

$$T_g^2 \frac{d^2}{dt^2} U_\omega + 2T_g \xi_g \frac{d}{dt} U_\omega + U_\omega = k_g \omega \tag{5.3}$$

Thus, gyro transfer functions can be obtained by Laplace transformation, and it is a typical second-order oscillation loop:

$$\frac{U_\omega(s)}{\omega(s)} = \frac{k_g}{T_g^2 s^2 + 2T_g \xi_g s + 1}, \tag{5.4}$$

where $\omega(s)$ is the angular rate sensed by the gyro, $U_\omega(s)$ is the gyro output, $\omega_g = \frac{1}{T_g}$ is the natural frequency of the second-order loop, ξ_g is relative damping coefficient, k_g represents the gain.

The attitude control system design has a clear requirement for k_g, ω_g, and ξ_g, i.e., the allowable deviations of k_g, ω_g, and ξ_g are generally 5, 10, and 30%, respectively.

This second-order model describes ideal gyroscope properties, but various influences of inertial component accuracy, installation error, overall damping, and output delay, etc., mean actual properties are not exactly the same as the model. Therefore, nonlinear characteristics should also be considered, such as relay characteristics, dead sensitivity area, saturation, clearance, and friction. In many cases, the gyro can be modeled as a linear transfer function with nonlinear input or output links, providing a mathematical model that more closely describes real characteristics.

Modern IMUs and rate gyros are digitized, where a digital filter is used to adjust dynamic characteristics. Therefore, features can be fitted depending on actual output, not limited to the second-order model, and linear and nonlinear characteristics can be combined.

5.3.1.2 Servo Mechanism

The OBC outputs attitude control commands to the servo mechanism to swing the engines, and LV stability can be achieved by altering the thrust direction.

The servos comprise many hydraulic and electrical components, and since they are the actuator of the control system, performance requirements include static and dynamic characteristics. Static characteristics include open-loop speed gain, actuator displacement or angle, swing angular rate, load torque, zero drift, etc. Dynamic characteristics include step and frequency response, etc. In the simulation, the servo mechanism can be described by a combination of transfer function and nonlinear characteristics.

5.3.1.3 Nonlinear Characteristics

Almost all onboard devices that affect the system performance have some nonlinear characteristics, which should be tested, fitted, and combined into the model. Figure 5.4 shows several typical nonlinear characteristics.

5.3.2 Launcher Motion

The 6-DOF LV model is a complex mathematical model and provides the basis for simulation. A detailed explanation is provided elsewhere [14–16], and the following discussion only describes the important aspects of the LV motion model used for simulations.

5.3.2.1 Coordinate System Transformation

Figure 5.5 shows the transformation from the launch inertial to the LV coordinate system, where φ, ψ, γ are the Euler angles of pitch, yaw, and roll, respectively Unless otherwise instructed, the Euler angles are rotated in the order $\varphi \rightarrow \psi \rightarrow \gamma$.

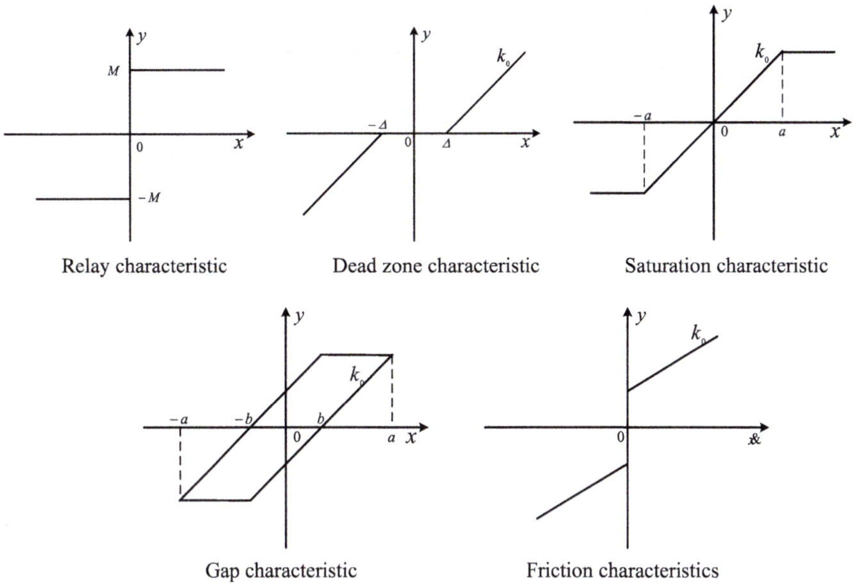

Fig. 5.4 Several typical nonlinear characteristics

Fig. 5.5 Transformation from the launch inertial to the LV coordinate system

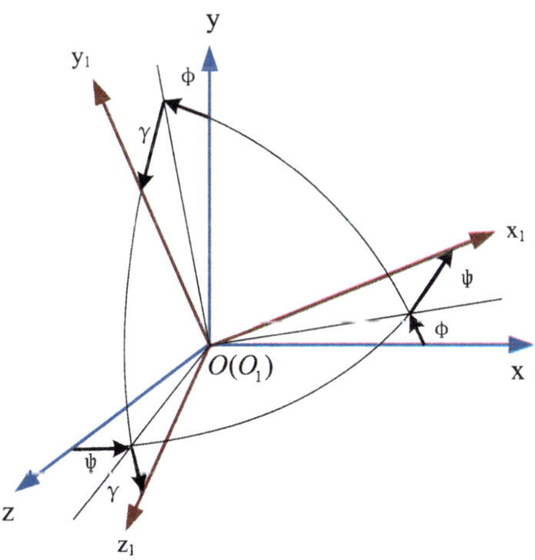

The transformation matrix from the LV to the launch inertial coordinate system is as follows:

$$\mathbf{A}=\begin{bmatrix} \cos\varphi\cos\psi & \cos\varphi\sin\psi\sin\gamma-\sin\varphi\cos\gamma & \sin\varphi\sin\gamma+\cos\varphi\sin\psi\cos\gamma \\ \sin\varphi\cos\psi & \cos\varphi\cos\gamma+\sin\varphi\sin\psi\sin\gamma & -\cos\varphi\sin\gamma+\sin\varphi\sin\psi\cos\gamma \\ -\sin\psi & \cos\psi\sin\gamma & \cos\psi\cos\gamma \end{bmatrix}. \tag{5.5}$$

5.3.2.2 Launcher Centroid Motion

The motion equation is a set of dynamic equations used to determine the rocket trajectory, which is related to the magnitude and direction of external forces and moments, including gravity, aerodynamic force and moment, thrust, control force and moment, etc. Large LVs also experience elastic and propellant sloshing motions, in addition to 6-DOF rigid body motion, but these have a negligible effect on the trajectory. Therefore, the LV is considered a rigid body for guidance control purposes.

Gravity

Newton's law of gravitation states that the gravitational force F between a particle of mass M and another particle of mass m is

$$F = f\frac{Mm}{r^2}, \tag{5.6}$$

where f is the universal gravitational constant, r represents the distance between the two mass points.

Assuming the Earth is spherical with uniform mass, the mass may be regarded as concentrated in the center, and the gravitational potential of a point r away from the center, U, is

$$U = \frac{fM}{r}, \tag{5.7}$$

where $\mu = fM$ is defined as the Earth gravitational coefficient.

However, the Earth is actually a heterogeneous sphere of complex shape, so the gravitational potential of the point can be expressed as

$$U = f\int_M \frac{\mathrm{d}m}{\rho}, \tag{5.8}$$

where ρ represents the distance between the Earth internal point with mass dm and the point outside the Earth.

It is very difficult to measure the Earth's surface and interior density to provide accurate gravitational force. To simplify the problem, the Earth is usually considered as spinning ellipsoid with uniform mass distribution, where the ellipsoid center coincides with Earth's center, the rotating axis coincides with Earth's rotation axis, and the ellipsoid mass is the same as the Earth's.

Considering the Earth J_2 coefficient, the gravitational potential expression for the ellipsoid Earth is

$$U = \frac{fM}{r}\left[1 + \frac{J_2}{2}\left(\frac{a_e}{r}\right)^2\left(1 - 3\sin^2\varphi_{\mathrm{dx}}\right)\right], \tag{5.9}$$

where a_e is the Earth equatorial radius, φ_{dx} is the geocentric latitude of the point in the space.

The gradient of the gravitational potential represents the gravitational acceleration vector for unit mass affected by the Earth's gravitational force,

$$\vec{g} = \mathrm{grad}U. \tag{5.10}$$

To facilitate calculation, gravitational acceleration is usually divided into components along the direction of r and in the Earth rotation direction, ω,

$$g_r = -\frac{fM}{r^2}\left[1 + \frac{3}{2}J_2\left(\frac{a_e}{r}\right)^2\left(1 - 5\sin^2\varphi_{dx}\right)\right], \tag{5.11}$$

$$g_\omega = -3\frac{fM}{r^2}J_2\left(\frac{a_e}{r}\right)^2\sin\varphi_{dx}. \tag{5.12}$$

Therefore, the component of the gravitational force in the inertial coordinate system, g_x, g_y, and g_z, respectively, can be expressed as

$$\begin{cases} g_x = g_r\frac{R_{0x}+x}{r} + g_\omega\frac{\omega_x}{\omega_e} \\ g_y = g_r\frac{R_{0y}+y}{r} + g_\omega\frac{\omega_y}{\omega_e} \\ g_z = g_r\frac{R_{0z}+z}{r} + g_\omega\frac{\omega_z}{\omega_e}, \end{cases} \tag{5.13}$$

where ω_e is the Earth rotation rate, ω_x, ω_y, and ω_z represent the components of ω_e in the inertial coordinate system, respectively; R_{0x}, R_{0y}, and R_{0z} represent the components of the distance from the coordinate origin to the Earth center in the inertial coordinate system, respectively; x, y, and z represent the point position in the inertial coordinate system, respectively.

Aerodynamic Forces and Moments

When the LV moves relative to air, pressure and friction are called aerodynamic forces, which are very important for flight control before the LV exits the atmosphere. However, LV shape is relatively symmetrical, so aerodynamic forces can be simplified as

$$\begin{aligned} X &= -C_x \cdot q \cdot S_M \\ Y &= C_y \cdot q \cdot S_M \\ Z &= -C_z \cdot q \cdot S_M, \end{aligned} \tag{5.14}$$

where X, Y, and Z are the drag, lift, and lateral force, respectively, from the atmosphere acting on the body, with directions along the three axes of the airflow velocity coordinate system. C_x, C_y, and C_z are the drag, lift, and lateral force coefficients, respectively, which are usually functions of Maher number, attack angle, and sideslip angle. S_M is the characteristic cross-sectional area of the body, and q is dynamic head;

$$q = \frac{1}{2} \cdot \rho \cdot V_w^2, \tag{5.15}$$

where V_w is the speed of an aircraft relative to the air flow, and ρ is atmospheric density.

The drag, lift, and lateral force and moment along LV coordinate system axes can be expressed as aerodynamic force

$$\begin{bmatrix} R_{x1} \\ R_{y1} \\ R_{z1} \end{bmatrix} = E \cdot \begin{bmatrix} X \\ Y \\ Z \end{bmatrix} \tag{5.16}$$

and aerodynamic moment

$$\begin{bmatrix} M_{Rx1} \\ M_{Ry1} \\ M_{Rz1} \end{bmatrix} = \begin{bmatrix} 0 \\ R_{z1} \cdot (X_d - X_Z) \\ -R_{y1} \cdot (X_d - X_Z) \end{bmatrix}, \tag{5.17}$$

where X_d is the distance from the LV aerodynamic pressure center to the theoretical tip of the LV body, which changes with the Maher number; X_Z is the distance from the LV center of mass to the theoretical tip, which changes with the mass of the launcher; and $\boldsymbol{E}$ is the transformation matrix from the velocity to LV coordinate systems, which is a function of attack and sideslip angle,

$$E = \begin{bmatrix} \cos\alpha\cos\beta & \sin\alpha\cos\beta & -\sin\beta \\ -\sin\alpha & \cos\alpha & 0 \\ \cos\alpha\sin\beta & \sin\alpha\sin\beta & \cos\beta \end{bmatrix}, \tag{5.18}$$

where α, β are the attack and sideslip angle, respectively.

The drag, lift, and lateral force coefficients; and the aerodynamic pressure center are all obtained from wind tunnel tests.

Engine Thrust, Control Force, and Torque

The active forces acting on the rocket are the engine thrust and the force generated by controlling the direction of the thrust vector. Steady engine thrust can be expressed as

$$P = P_{b0}\dot{m}g_0 + S_a p_0 \left(1 - \frac{p}{p_0}\right), \tag{5.19}$$

where P_{b0} is the engine's ground-specific impulse, $\dot{m}$ is the fuel consumption per second, S_a is the engine ejector section area, p_0 is the ground standard atmospheric pressure, p is instant atmospheric pressure, $S_a p_0\left(1-\frac{p}{p_0}\right)$ is the static pressure difference.

If the LV has left the atmosphere, the thrust can be calculated from the product of the vacuum specific impulse and fuel consumption rate, and static pressure differences need not be considered.

Control force and torque are usually produced by gas vane deflection, engine swing, attitude control nozzles. Engine swing is widely adopted during the main engine working phase, where servos swing the engine nozzles to generate transverse and normal forces, then controlling torques around the three body axes to control pitch, yaw, and roll attitudes individually.

Figure 5.6 shows the engine identifiers and swing angles for the case where the engines are configured in "+" format. Assuming each engine's thrust is P_1, and the swing angles are δ_1, δ_2, δ_3, and δ_4 for the four engines, respectively, the forces along the three LV body axes can be expressed as

$$\begin{cases} P_{x1} = 4P_1 - P_1(\cos\delta_1 + \cos\delta_2 + \cos\delta_3 + \cos\delta_4) \\ P_{y1} = -P(\sin\delta_2 + \sin\delta_4) \\ P_{z1} = -P(\sin\delta_1 + \sin\delta_3) \end{cases} \tag{5.20}$$

and the corresponding torques as

$$\begin{cases} M_{Px1} = -P_1 R_c(-\sin\delta_1 + \sin\delta_2 + \sin\delta_3 - \sin\delta_4) \\ M_{Py1} = -P_1(\sin\delta_1 + \sin\delta_3)(X_c - X_Z) \\ M_{Pz1} = P_1(\sin\delta_2 + \sin\delta_4)(X_c - X_Z), \end{cases} \tag{5.21}$$

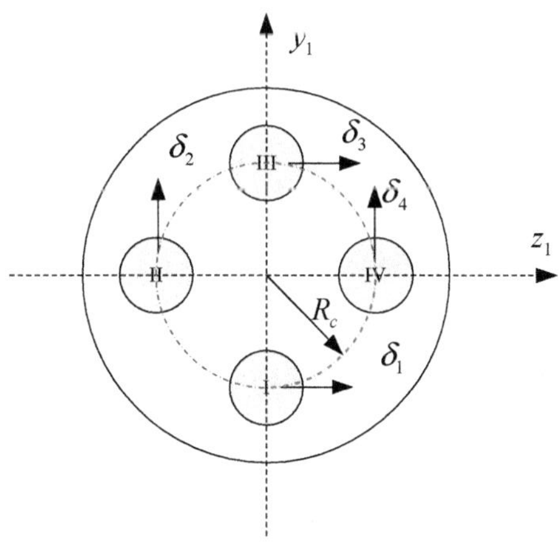

Fig. 5.6 "+" shaped engine swing angle diagram

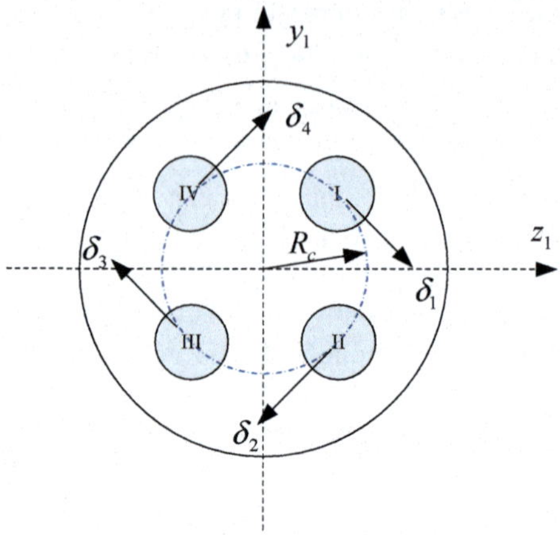

Fig. 5.7 "X" shaped engine swing angle diagram

where the R_c represents the distance from the engine ejector center to the LV longitudinal axis, X_c is the distance from the projecting point of the engine ejector center on the LV longitudinal axis to the theoretical tip of the LV body.

Figure 5.7 shows the engine identifiers and swing angles for the case where the engines are configured in "X" format. Using the same assumptions as the previous case, the forces along LV body axes can be expressed as

$$\begin{cases} P_{x1} = 4P_1 - P_1(\cos\delta_1 + \cos\delta_2 + \cos\delta_3 + \cos\delta_4) \\ P_{y1} = P_1\cos 45^\circ(\sin\delta_1 + \sin\delta_2 - \sin\delta_3 - \sin\delta_4) \\ P_{z1} = P_1\cos 45^\circ(\sin\delta_1 - \sin\delta_2 - \sin\delta_3 + \sin\delta_4), \end{cases} \tag{5.22}$$

and the corresponding torques as

$$\begin{cases} M_{Px1} = -P_1 R_c(\sin\delta_1 + \sin\delta_2 + \sin\delta_3 + \sin\delta_4) \\ M_{Py1} = -P_1\cos 45^\circ(\sin\delta_1 - \sin\delta_2 - \sin\delta_3 + \sin\delta_4)(X_c - X_Z) \\ M_{Pz1} = -P_1\cos 45^\circ(\sin\delta_1 + \sin\delta_2 - \sin\delta_3 - \sin\delta_4)(X_c - X_Z). \end{cases} \tag{5.23}$$

Launcher Motion

The launcher motion equations are based on the general motion laws of a rigid body with variable mass and mass point. The trajectory equation is derived in the launch point relative coordinate system, which rotates with the Earth; and the launch point inertial coordinate system, which can be regarded as fixed at liftoff, and then becomes an inertial coordinate system with respect to the Earth's gravitational field.

Free rigid body motion in space can be decomposed into mass center motion and rotation about the mass center, i.e., 6-DOF motion. The motion equations can be expressed in an inertial coordinate system as

$$\begin{cases} \dot{V}_x = a_x = \frac{P_x}{m} + \frac{R_x}{m} + g_x \\ \dot{V}_y = a_y = \frac{P_y}{m} + \frac{R_y}{m} + g_y \\ \dot{V}_z = a_z = \frac{P_z}{m} + \frac{R_z}{m} + g_z \end{cases}. \tag{5.24}$$

$$\begin{cases} \dot{x} = V_x \\ \dot{y} = V_y \\ \dot{z} = V_z \end{cases}. \tag{5.25}$$

$$\begin{cases} J_{x1}\dot{\omega}_{x1} - (J_{y1} - J_{z1})\omega_{y1}\omega_{z1} = M_{Px1} + M_{Rx1} \\ J_{y1}\dot{\omega}_{y1} - (J_{z1} - J_{x1})\omega_{z1}\omega_{x1} = M_{Py1} + M_{Ry1} \\ J_{z1}\dot{\omega}_{z1} - (J_{x1} - J_{y1})\omega_{x1}\omega_{y1} = M_{Pz1} + M_{Rz1}, \end{cases} \tag{5.26}$$

where *V*, *a*, *P*, *R*, and *g* represent the velocity, acceleration, thrust vector, aerodynamic force, and gravity, respectively; suffix *x*, *y*, and *z* represent the components in the inertial coordinate system, respectively. *J* and ω represent the moments of inertia and the angular rates around the measurement axes, respectively; suffix *x1*, *y1*, and *z1* represent the components in the LV coordinate system, respectively, *m* is the LV mass.

$$\begin{bmatrix} P_x \\ P_y \\ P_z \end{bmatrix} = \boldsymbol{A} \begin{bmatrix} P_{x1} \\ P_{y1} \\ P_{z1} \end{bmatrix}, \tag{5.27}$$

$$\begin{bmatrix} R_x \\ R_y \\ R_z \end{bmatrix} = \boldsymbol{A} \begin{bmatrix} R_{x1} \\ R_{y1} \\ R_{z1} \end{bmatrix}, \tag{5.28}$$

where $\boldsymbol{A}$ is the transformation matrix from the LV to the inertial coordinate system (see Eq. 5.5).

However, this does not consider various disturbance factors. Many parameters are not completely identical in real conditions with the theoretical value, and random disturbances will have a certain probability and amplitude. To facilitate data analysis, disturbances and deviations are usually extracted as part of the model.

5.3.2.3 Rotation Around the Launcher Mass Center

Rotation around the launcher mass center can be modeled in the LV coordination system as

$$\begin{bmatrix}\dot{\omega}_{x1}\\ \dot{\omega}_{y1}\\ \dot{\omega}_{z1}\end{bmatrix} = \begin{bmatrix}\omega_{y1}\omega_{z1}(J_{y1}-J_{z1})/J_{x1}\\ \omega_{x1}\omega_{z1}(J_{z1}-J_{x1})/J_{y1}\\ \omega_{x1}\omega_{y1}(J_{x1}-J_{y1})/J_{z1}\end{bmatrix} + \begin{bmatrix}(M_{Px1}+M_{Rx1})/J_{x1}\\ (M_{Py1}+M_{Ry1})/J_{y1}\\ (M_{Pz1}+M_{Rz1})/J_{z1}\end{bmatrix}. \tag{5.29}$$

The relationship between the Euler angles and the angular rates around the LV coordination system measurement axes is as follows:

$$\begin{bmatrix}\dot{\gamma}\\ \dot{\psi}\\ \dot{\varphi}\end{bmatrix} = \begin{bmatrix}1 & \tan\psi\sin\gamma & \tan\psi\cos\gamma\\ 0 & \cos\gamma & -\sin\gamma\\ 0 & \sin\gamma/\cos\psi & \cos\gamma/\cos\psi\end{bmatrix}\begin{bmatrix}\omega_{x_1}\\ \omega_{y_1}\\ \omega_{z_1}\end{bmatrix}, \tag{5.30}$$

where φ, ψ, and γ are the pitch, yaw, and roll attitude angles, respectively.

Euler angle deviations are as follows:

$$\begin{bmatrix}\Delta\gamma\\ \Delta\psi\\ \Delta\varphi\end{bmatrix} = \begin{bmatrix}\gamma-\gamma_{cx}\\ \psi-\psi_{cx}\\ \varphi-\varphi_{cx}\end{bmatrix}, \tag{5.31}$$

where "Δ" represents the deviation, φ_{cx}, ψ_{cx}, and γ_{cx} represent the pitch, yaw, and roll program angles, respectively.

Attitude angle deviations in the LV coordinate system are as follows:

$$\begin{bmatrix}\Delta\gamma_{x1}\\ \Delta\psi_{y1}\\ \Delta\varphi_{z1}\end{bmatrix} = \begin{bmatrix}1 & 0 & -\sin\psi\\ 0 & \cos\gamma & \cos\psi\sin\gamma\\ 0 & -\sin\gamma & \cos\psi\cos\gamma\end{bmatrix}\begin{bmatrix}\Delta\gamma\\ \Delta\psi\\ \Delta\varphi\end{bmatrix}, \tag{5.32}$$

where $\Delta\varphi_{z1}$, $\Delta\psi_{y1}$, and $\Delta\gamma_{x1}$ are the attitude angle deviations, respectively, around the LV coordinate system measurement axes.

5.3.2.4 Launcher Small Deviation Motion

Generally, LV attitude motion equations in pitch, yaw, and roll channels can be expressed as Eqs. (5.33)—(5.35), and due to the launcher's axisymmetric structure, the small deviation equation coefficients of the pitch and yaw channels are the same.

For pitch channel

$$\begin{cases}\Delta\dot{\theta} = c_1\Delta\alpha + c_2\Delta\theta + c_3\Delta\delta_\varphi + c_4\Delta\dot{\varphi} + c_3''\Delta\ddot{\delta}_\varphi + c_1'\alpha_w\\ \quad + \sum c_{4p}\Delta\ddot{y}_p + \sum(c_{1i}\dot{q}_{iy} + c_{2i}q_{iy}) + \bar{F}_{byc}\\ \Delta\ddot{\varphi} + b_1\Delta\dot{\varphi} + b_2(\Delta\alpha + \Delta\alpha_w) + b_3\delta_\varphi + b_3''\Delta\ddot{\delta}_\varphi + \sum(b_{1i}\dot{q}_{iy0} + b_{2i}q_{iy})\\ \quad + \sum(b_{4p}\Delta\ddot{y}_p - b_{5p}\Delta y_p) = \bar{M}_{bz1}.\\ \Delta\theta = \Delta\varphi - \Delta\alpha\end{cases} \tag{5.33}$$

For yaw channel

$$\begin{cases} \dot{\sigma} = c_1\beta + c_2\sigma + c_3\delta_\psi + c_3''\ddot{\delta}_\psi + c_4\dot{\psi} + c_1'\beta_w \\ \quad + \sum c_{4p}\Delta\ddot{z}_p + \sum (c_{2i}q_{iz} + c_{1i}\dot{q}_{iz}) + \bar{F}_{bzc} \\ \ddot{\psi} + b_1\dot{\psi} + b_2(\beta + \beta_w) + b_3\delta_\psi + b_3''\Delta\ddot{\delta}_\psi + \sum (b_{1i}\dot{q}_{iz} + b_{2i}q_{iz}) \\ \quad + \sum (b_{4p}\ddot{z}_p - b_{5p}z_p) = \bar{M}_{by1}. \\ \sigma = \psi - \beta \end{cases} \tag{5.34}$$

For roll channel

$$\ddot{\gamma} + d_1\dot{\gamma} + d_3\delta_\gamma + d_3''\ddot{\delta}_\gamma + \sum (d_{1i}\dot{q}_{in} + d_{2i}q_{in}) = \bar{M}_{bx1} \tag{5.35}$$

where $\Delta\theta$ is the ballistic tilt angle deviation, σ is the course angle; c_1 is the aerodynamic coefficient, c_1' is the aerodynamic damping force coefficient, c_2 is the coefficient associated with the gravity, c_3 is the control force coefficient, c_3' is the engine swinging inertial force coefficient, c_4 is the transverse and normal aerodynamic damping coefficient; b_1 is the aerodynamic damping moment coefficient, b_2 is the aerodynamic stability moment coefficient, b_3 is the control torque coefficient, b_3'' is the engine swinging acceleration inertial moment coefficient of pitch and yaw channels; d_1 is the rolling damping moment coefficient, d_3 is the rolling control torque coefficient, d_3'' is the engine swinging acceleration inertial moment coefficient of roll channel. δ_φ, δ_ψ, and δ_γ are the engine swing angles for pitch, yaw, and roll channel, respectively. α_w is the wind attack angle caused by wind disturbance, β_w is the sideslip angle caused by wind disturbance. $\overline{F}_{byc}$ is the synthetic interference for pitch channel, $\overline{F}_{bzc}$ is the synthetic interference for yaw channel. c_{4p} is the sloshing coupling force coefficient of pitch and yaw channels, b_{4p} is the sloshing coupling moment coefficient of pitch and yaw channels, b_{5p} is the sloshing centroid coupling coefficient of pitch and yaw channels. c_{1i} is the elastic additional aerodynamic damping force coefficient of the pitch and yaw channels, c_{2i} is the elastic additional aerodynamic force and additional thrust coefficient of the pitch and yaw channels, b_{1i} is the is the elastic additional aerodynamic damping moment coefficient of the pitch and yaw channels, b_{2i} is the elastic additional aerodynamic moment and additional thrust moment coefficient of the pitch and yaw channels. d_{1i} is the elastic aerodynamic damping moment coefficient of roll channel, d_{2i} is the elastic moment coefficient of the roll channel. $\overline{M}_{bz1}$, $\overline{M}_{by1}$, and $\overline{M}_{bx1}$ are the synthetic interference moment coefficient for pitch, yaw, and roll channel, respectively. $\ddot{y}_p$, y_p, and $\ddot{z}_p$, z_p are acceleration and displacement, respectively, for normal and transverse sloshing of a propellant mass point in the storage tank, and suffix p is the storage tank indicator. q_{iy}, $\dot{q}_{iy}$ and q_{iz}, $\dot{q}_{iz}$ are generalized coordinates and velocities, respectively, of the ith-order normal and transverse LV elastic vibration, with velocity in generalized coordinates. q_{in} and $\dot{q}_{in}$ are the generalized coordinates and velocities of the ith-order torsional vibration of the launcher, and the velocity in the generalized coordinates.

The small deviation equation coefficients can be calculated according to the standard trajectory, and the interpolated parameters for the coefficients can be applied for small deviation-state simulations, i.e., simulation and nominal state differences are limited within a specific scope.

5.3.2.5 Launcher Elastic Motion

The launcher elastic motion equations can be expressed as

$$\begin{cases} \ddot{q}_{iy} + 2\xi_i\omega_i\dot{q}_{iy} + \omega_i^2 q_{iy} = D_{1i}\omega_{z1} + D_{2i}(\alpha + \alpha_w) + D_{3i}\delta_\varphi + D''_{3i}\ddot{\delta}_\varphi \\ \quad + \sum\left(G''_{ip}\ddot{y}_p + G_{ip}y_p\right) - Q_{iy} \\ \ddot{q}_{iz} + 2\xi_i\omega_i\dot{q}_{iz} + \omega_i^2 q_{iz} = D_{1i}\omega_{y1} + D_{2i}(\beta + \beta_w) + D_{3i}\delta_\psi + D''_{3i}\ddot{\delta}_\psi \\ \quad + \sum\left(G''_{ip}\ddot{z}_p + G_{ip}z_p\right) - Q_{iz} \\ \ddot{q}_{in} + 2\xi_{in}\omega_{in}\dot{q}_{in} + \omega_i^2 q_{iz} = d_{31in}\delta_\psi + d''_{31in}\ddot{\delta}_\gamma - Q_{in}, \end{cases} \quad (5.36)$$

where ω_i is the ith vibration mode natural frequency, ω_{in} is the ith-order torsional vibration mode natural frequency, ξ_i is the ith-order vibration mode damping coefficient, ξ_{in} is the ith-order torsional vibration mode damping coefficient, D_{1i} and D_{2i} are the effect of the angular velocity and the attack angle equivalent generalized force, respectively, on the vibration of the ith-order vibration mode; D_{3i}, and D''_{3i} are the influence coefficients of the engine swing angle and angular acceleration, respectively, on the vibration of the ith-order vibration mode; d_{31in}, and d''_{31in} are the influence coefficients of the engine swing angle and angular acceleration, respectively, on the vibration of the ith-order torsional vibration mode; G_{ip}, and G''_{ip} are the influence coefficients of the sloshing generalized coordinate and acceleration in storage bank, respectively, on the vibration of the ith-order vibration mode, with acceleration in generalized coordinate; Q_{iy}, and Q_{iz} are the normal and transverse generalized forces corresponding to the ith-order vibration mode; Q_{in} is the generalized force corresponding to the ith-order torsional vibration mode. $\dot{\omega}_{z1}$ and $\dot{\omega}_{y1}$ are the angular accelerations around the centroid in the 6-DOF simulation, respectively, which can be simplified as $\Delta\dot{\varphi}$ and $\dot{\psi}$, respectively, in the launcher small deviation motion equations. The definitions of other variables are shown in Eqs. (5.33)–(5.35).

5.3.2.6 Launcher Sloshing Motion

The launcher sloshing motion equations can be expressed as

$$\begin{cases} \ddot{y}_p + 2\xi_p\Omega_p\dot{y}_p + \Omega_p^2 y_p = -E_{pz}\dot{\omega}_{z1} - a_{yb} + \sum\left(E''_{ip}\ddot{q}_{iy} + E_{ip}q_{iy}\right) \\ \ddot{z}_p + 2\xi_p\Omega_p\dot{z}_p + \Omega_p^2 z_p = -E_{pz}\dot{\omega}_{y1} + a_{zb} + \sum\left(E''_{ip}\ddot{q}_{iz} + E_{ip}q_{iz}\right) \end{cases}, \quad (5.37)$$

where Ω_p is the natural frequency of liquid propellant sloshing in the storage tank, ξ_p is the sloshing damping coefficient of the storage tank, E_{pz} is the influence coefficient of the angular acceleration on the sloshing, with angular acceleration around centroid; a_{yb} and a_{zb} are the apparent acceleration in the LV coordinate system, and $a_{yb} = \dot{W}_{y1}$, $a_{zb} = \dot{W}_{z1}$; E_{ip} and E''_{ip} are the liquid sloshing elastic coupling coefficient on the ith-order generalized coordinate and acceleration. The definitions of other variables are shown in Eqs. (5.33)–(5.36).

5.3.2.7 Launcher 6-DOF Simulation

The LV 6-DOF simulation model can be described as

$$\begin{cases} \dot{V}_x = \dot{W}_x + g_x \\ \dot{V}_y = \dot{W}_y + g_y \\ \dot{V}_z = \dot{W}_z + g_z \end{cases}, \tag{5.38}$$

$$\begin{cases} \dot{x} = V_x \\ \dot{y} = V_y \\ \dot{z} = V_z \end{cases}, \tag{5.39}$$

where V, W, and g represent the velocity, apparent velocity, and gravity, respectively, in the inertial coordinate system; x, y, and z represent the position, respectively, in the inertial coordinate system; suffix x, y, and z represent the components in the inertial coordinate system.

The apparent acceleration in the LV coordinate system can be expressed as

$$\begin{cases} \dot{W}_{x1} = F_{x1}/m \\ \dot{W}_{y1} = F_{y1}/m \\ \dot{W}_{z1} = F_{z1}/m \end{cases}, \tag{5.40}$$

and

$$\begin{cases} F_{x1} = P_{x1} + R_{x1} \\ F_{y1} = P_{y1} + R_{y1} \\ F_{z1} = P_{z1} + R_{z1} \end{cases}, \tag{5.41}$$

where F, P, R represent the external force, engine thrust, and aerodynamic force, respectively, acting on the LV, suffix $x1$, $y1$, and $z1$ represent the components in the LV coordinate system. m is the mass of the LV.

The apparent acceleration in the LV coordinate system is converted to the inertial coordinate system

$$\begin{bmatrix} \dot{W}_x \\ \dot{W}_y \\ \dot{W}_z \end{bmatrix} = \boldsymbol{A} \begin{bmatrix} \dot{W}_{x1} \\ \dot{W}_{y1} \\ \dot{W}_{z1} \end{bmatrix}, \tag{5.42}$$

where $\boldsymbol{A}$ is the transformation matrix.

The motion equations rotating around the centroid with elastic vibration and propellant sloshing included are expressed as follows, in the LV coordinate system

$$\begin{bmatrix} \dot{\omega}_{x1} \\ \dot{\omega}_{y1} \\ \dot{\omega}_{z1} \end{bmatrix} = \begin{bmatrix} \omega_{y1}\omega_{z1}(J_{y1} - J_{z1})/J_{x1} \\ \omega_{x1}\omega_{z1}(J_{z1} - J_{x1})/J_{y1} \\ \omega_{x1}\omega_{y1}(J_{x1} - J_{y1})/J_{z1} \end{bmatrix} - \begin{bmatrix} \bar{M}_{\ddot{\delta}_\gamma} \\ \bar{M}_{\ddot{\delta}_\psi} \\ \bar{M}_{\ddot{\delta}_\phi} \end{bmatrix} - \begin{bmatrix} 0 \\ \sum (b_{4p}\ddot{z}_p - b_{5p}z_p) \\ \sum (b_{4p}\ddot{y}_p - b_{5p}y_p) \end{bmatrix} - \begin{bmatrix} \sum (d_{1i}\dot{q}_{in} + d_{2i}q_{in}) \\ \sum (b_{1i}\dot{q}_{iz} + b_{2i}q_{iz}) \\ \sum (b_{1i}\dot{q}_{iy} + b_{2i}q_{iy}) \end{bmatrix} + \begin{bmatrix} (M_{Px1} + M_{Rx1})/J_{x1} \\ (M_{Py1} + M_{Ry1})/J_{y1} \\ (M_{Pz1} + M_{Rz1})/J_{z1} \end{bmatrix}, \tag{5.43}$$

where $\bar{M}_{\ddot{\delta}_\gamma}$, $\bar{M}_{\ddot{\delta}_\psi}$, and $\bar{M}_{\ddot{\delta}_\varphi}$ are the moments produced by engine swing. The definitions of other variables are shown in Eqs. (5.26), (5.33)–(5.37).

Equation (5.30) shows the relationship between the Euler angle and angular motion around the LV coordinate system, and Eq. (5.31) shows the Euler deviation, and attitude angle deviation around the LV axes, where $\Delta\varphi_{z1}$, $\Delta\psi_{y1}$, $\Delta\gamma_{x1}$ are the attitude angle deviations in the LV coordinate systems, respectively, and used for LV control.

Since the LV usually flies in the target plane, $\psi_{cx} = 0$ and $\gamma_{cx} = 0$, and hence $\Delta\psi = \psi$, $\Delta\gamma = \gamma$. The small deviation motion introduced in Sect. 5.3.2.4 refers to the deviation from the standard trajectory, hence the attitude angle motion equation consists of $\Delta\varphi$, ψ, and γ.

5.4 Simulation Process

5.4.1 DOF Model Calculation

Modeling design inputs include mass, inertia, engine thrust model, aerodynamics, elasticity, sloshing, wind, engine angle synthesis and decomposition, launch point, target point, and related parameters, and each device's characteristic parameters. Each device model should be set up and initialized based on the relevant data, parameters, and details.

Figure 5.8 shows the trigger sequence for the circulatory model calculation. The swing command is calculated based on control module input. Then the servo model

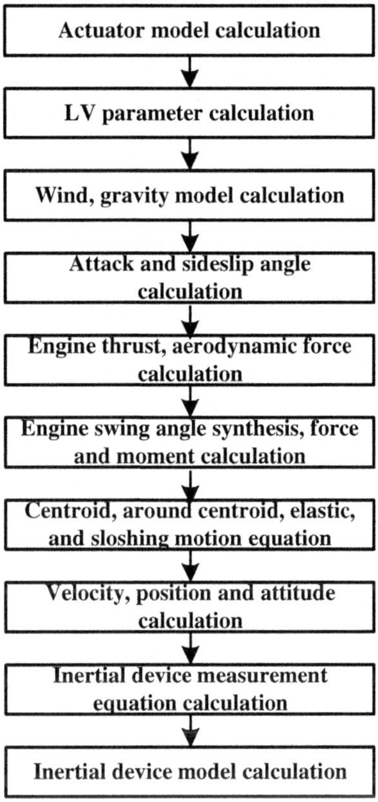

Fig. 5.8 6-DOF model calculation flow

provides actual engine swing angles for the current control cycle. Launch vehicle parameters, and the wind and gravity field models are calculated depending on flight elevation, as well as attack and sideslip angles. Engine thrust, aerodynamic force, and various torques are then calculated, and based on the calculation results, the numerical integral of centroid, around centroid, elastic, and sloshing motion equations are conducted. Runge–Kutta or Adams methods are usually adopted for the numerical integration. Finally, measurements that the inertial device should sense are calculated.

For HWIL simulations, inertial device and actuator (servo) models are replaced by the real prototypes, and the corresponding model calculations can be omitted.

5.4.2 Control Module Calculation

Figure 5.9 shows that control module calculations generally commence with inertial measurement data treatments. Inertial measurement data are sampled in each control cycle, including IMU, rate gyros, and accelerometer assemblies. If these are

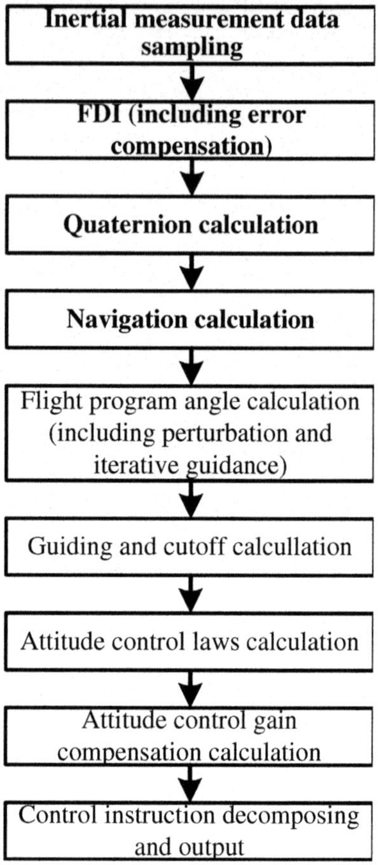

Fig. 5.9 Onboard control module calculation flow

redundant configured, FDI is performed first, otherwise, the FSW directly applies error compensation for inertial data, then calculates the quaternions to obtain LV flight attitude angle, position, and velocity. The real-time flight program angle is then determined based on the guidance laws, and guide signals, and cutoff equation are calculated.

Attitude control computation then commences. Attitude angle deviation, angular rate, and transverse and normal accelerations are passed to the control laws, to generate attitude control signals. These signals are decomposed depending on the engine configurations, converted to control instructions, and sent to the actuators (or servos).

5.5 Attitude Control Simulation States

Attitude control simulation focuses on stability and design margins, forms the basis for the 6-DOF simulation test, and includes the following simulation states.

5.5.1 Deviation

Deviation simulation is the key aspect of testing. The rated condition simulation is first set, i.e., LV model parameters are taken as their theoretical or rating values. Then deviation simulation is performed with upper and lower bound states for each parameter being checked, or a combination. This evaluates design margins and the attitude control adaptability. Upper and lower limit states, which exceed upper and lower bound states, are also checked, i.e., parameter deviations reach the limit boundaries, which is equivalent to increasing the static gain bias.

Table 5.1 shows an example where the parameter deviations relative to the rating depend on engineering experience, model parameter accuracy, design margins, product performance, manufacturing processes, etc.

5.5.2 Load Reduction

Active load reduction is simulated to evaluate the design effect. Wind disturbances are considered independently in the motion equations, with wind shear disturbances α_{wq} and β_{wq} considered within 5–20 km LV elevation. During LV first-stage flight,

Table 5.1 Attitude control simulation parameters deviation combinations

Parameter name	Upper bound (%)	Lower bound (%)
System gain	+20	-20
Guiding gain	+15	+15
Control moment coefficient	+20	−20
Control force coefficient	+20	−20
Rolling control moment coefficient	+15	−15
Sloshing frequency	−10	−10
Sloshing force coupling coefficient	+10	−10
Sloshing moment coupling coefficient	+15	−15
Sloshing centroid coupling coefficient	+10	−10
Elastic vibration damping coefficient	20	−20
Elastic vibration frequency	−15 ~ −30	+15 ~ +30%
Elastic vibration mode slope where the IMU is installed	Same polarity, 10 ~ 12	Same polarity, 10 ~ 12
Elastic vibration mode slope where the rate gyro is installed	Same polarity, 10 ~ 12	Same polarity, 10 ~ 12
Elastic coupling coefficient	+25	−25
Torsional vibration damping coefficient	−20	−20
Torsional vibration frequency	−30 ~ −50	+30 ~ +50
Torsional vibration mode where the IMU is installed	Same polarity, 10 ~ 15	Same polarity, 10 ~ 15
Torsion coupling coefficient	+30	−30

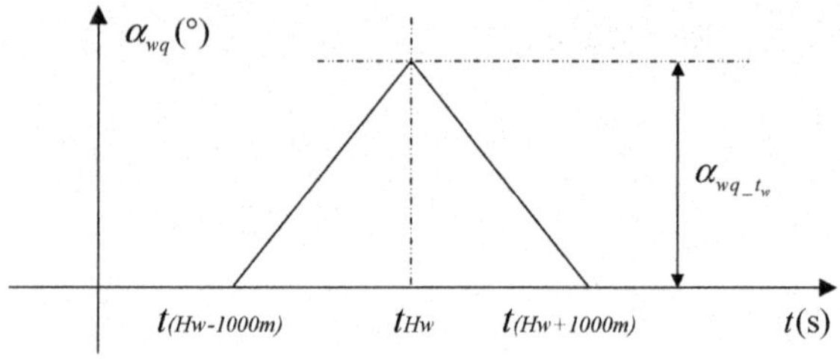

Fig. 5.10 Wind shear diagrammatic sketch

the moment t_{Hw}, when wind shear is introduced into the pitch and yaw control channel can be set at the time attaining transonic velocity, $|b_2|_{\max}$, maximum dynamic head, etc. Figure 5.10 shows the wind shear diagram, and the action scope is considered as 2000 m.

5.5.3 Other Normal Simulation Conditions

1. **Initial conditions**

Stability of key points, such as takeoff, booster separation, second-stage control start-up, fairing separation, etc., should be checked depending on the various initial conditions, i.e., different parameters' initial settings should be simulated to check attitude control adaptability.

2. **Disturbances**

Stability in the presence of control loop disturbances should be checked. Disturbances are usually introduced in the attitude control instruction power amplification output end, or servo mechanism swing angle feedback. These analog signals can be easily interfered, and the electromagnetic environment is also poor. Disturbances can be simulated as sine and cosine signals with different frequencies and amplitudes.

3. **System phase margin**

The system phase margin should be tested and evaluated by increasing system delay

4. **Nonlinear characteristics**

Nonlinear characteristics testing mainly relate to the servo mechanisms. A small angle sinusoidal signal is stimulated and sent to the servo mechanism under engine load, to check the actuator's dynamic characteristics from the angular displacement sensor outputs. Thus, nonlinear characteristics, such as dead zones, are recorded and analyzed.

5.5.4 *Fault Simulation Test*

The fault simulation test checks redundant function, and failure effects on the attitude control system. Typical simulated failure modes include rate gyro failure (zero, saturation, unchanging, or random outputs), transverse and normal accelerometer assembly failure (zero, unchanging, etc.), IMU failure, servos failure (no swing, stuck, zero or saturation output, open loop, etc.), thrust drop, attitude control stage transition failure (ahead of schedule, or lag schedule), etc. Fault timing can be determined by theoretical analysis, usually when the failure has the most influence on system performance, or minimum control margins, maximum load, or the most severe state fluctuation occur.

Simulated IMU failures include gyros and accelerometers. During the attitude control simulation, only fault consequences are considered, rather than the specific hardware and software failure mechanism. Typical failure consequences include no, saturated, constant, output,etc. This redundancy testing is combined with guidance system testing, and rationality of the fault thresholds can be verified.

For the iterative guidance method, if the thrust drop does not affect controllability, and propellant is not exhausted, the guidance method can still ensure accurate injection. Therefore, the emphasis of the attitude control simulation under thrust drop is to check if the flight remains stable. Timing, magnitude, and duration of thrust drop; and thrust recovery details, such as timing, proportion recovered, etc., can be set up to assess control system performance.

References

1. Bao WM (2013) The development trend of aerospace craft simulation technology. Aerospace Control 31(2):4–8 (in Chinese)
2. Eickhoff J (2009) Simulating spacecraft systems. Springer-Verlag, Berlin
3. Wang HL, Cao JG et al (2003) Design and application of simulation system. Science Press, Beijing. (in Chinese)
4. Xu YW (editor in chief) (ed) (1992) Control system (3). Missile and aerospace collection—liquid ballistic missile and launch vehicle series. China Aerospace Publishing House, Beijing. (in Chinese)
5. Cervantes D, Montaez L, Tatge L (2004) Low cost test bed tool development for validation of mission critical events. In: IEEE Aerospace Conference Proceedings, Big Sky, MT, USA, March 6–13
6. Cockrell CE, Taylor JL, Tuma M et al (2006) Integrated system test approaches for the NASA Ares I crew launch vehicle. In: 57th international astronautical congress. Valencia, Spain, October 2–6
7. Hughes R, Walker D (2009) ARTEMIS: Ares real time environment for modeling, integration, and simulation. In: AIAA modeling & Simulation Technologies Conference, Chicago, Illinois, USA, August 10–13
8. Kromis PA Modeling and simulation of the ARES UPPER STAGE transportation, lifting, stacking and mating operations within the vehicle assembly building at KSC[DB/OL]. http://ntrs.nasa.gov/archive/nasa/casi.ntrs.nasa.gov/20100042551.pdf

9. Li L (2014) Research on the application of the CZ-7 launch vehicle digital joint operation simulation technology. In: 3rd academic conference on manned spaceflight, Chengdu, Nov. 13–15. (in Chinese)
10. Slagowski SE, Vican JE (2007) A hardware-in-the-loop simulator for software development for a Mars airplane. In: AIAA modeling and simulation technologies conference and exhibit, Hilton Head, South Caroline, USA, August 20–23
11. Milenkovic Z, D'Souza C (2012) The space operations simulation center (SOSC) and closed-loop hardware testing for Orion rendezvous system design. In: AIAA guidance, navigation, and control conference, Minneapolis, Minnesota, USA, August 13–16
12. Liu LJ, Xu GB (1998) Study on six degree of freedom digital simulation of launch vehicle control system. Aerospace Control 16(3):46–52, 45. (in Chinese)
13. Deng FL, Yan B, Wang SC (1999) The research and expectation of six-degree-of-freedom simulation of large scale ballistic vehicle. J Astronautics 20(3):106–111. (in Chinese)
14. Zipfel PH (2007) Modeling and simulation of aerospace vehicle dynamicS, 2nd edition. Reston, AIAA Education Series
15. Wie B (1998) Space vehicle dynamics and control. Arizona state University, AIAA Education Series
16. Du W (2010) Dynamic modeling and ascent flight control of Ares-1 Crew Launch Vehicle. Ames, Iowa State University

Chapter 6
Launch Control

The TLCS [1] has two major components. Testing aspects were introduced and discussed above, and this chapter concentrates on launch control. Launch control is critical for successful LV flight and includes generating launch elements (data) and software, uploading flight software, determining initial state, ignition control, and implementing emergency shutdown after a failure.

Many countries have recognized TLCS importance to address increasing competition in the LV market. For example, the NASA space technology development roadmap in 2010 and 2015 had ground and launch system processing as the 13th area of concern, where the technologies focusing with TLCS include automatic command and control, advanced launch technology, fault diagnosis, isolation, etc. The 2010 document also showed a variety of ground operations (not only TLCS) accounted for 40% of the project total cost, hence simplifying operation and improving efficiency has significant potential to reduce the cost to reach "ship and shoot" target. Nowadays, onboard autonomous checkout, ground intelligent data analysis, simplified manual operation and ground equipment, and reduced technical support personnel are becoming more important.

Launch control systems usually adopt front-end and back-end configurations [2, 3] to ensure safety, where front-end equipments are installed in the basement or the launch pad, and back-end equipment in a control center, with the two systems connected by an optical fiber network. Back-end equipments are mainly computers forwarding instructions issued by the operators through the network to the front-end. Front-end equipments are actuators, converting the instructions into actual operations, and feeding back operational results to the back-end for display or further analysis.

This chapter briefly introduces different LV launch control systems (LCSs), expanding and supplementing Chap. 1, including the Chinese LV prelaunch process, and the instantaneous launch window in particular. Then LCS composition

Z. Song, *The Test and Launch Control Technology for Launch Vehicles*,
https://doi.org/10.1007/978-981-10-8712-7_6

is discussed, including the front-end and back-end systems, communication network, ground launch control software, etc. Launch control software has significantly higher safety requirements than other ground software, hence software reliability design, software testing, and special launch control software technology, such as process control language (PCL), are also introduced. With increasing application of cryogenic engines, engine fueling, pressure control, and prelaunch testing and preparation have been extremely complicated with correspondingly heavy workloads. However, the required technologies are similar to traditional LCS and are not discussed in this chapter.

6.1 Introduction

6.1.1 US Launch Control System

In NASA's definition, a space project architecture includes ground, mission, LV, spacecraft, etc. systems. The GS also includes many elements, such as spacecraft processing, recovery, and retrieval; LV processing; mobile launch; launch pad; command, control, and communication; and operations support. Each element consists of multiple subsystems, where the command, control, and communication system includes launch control, imagery, and communication systems, and control center facility [4].

Although the US launch control system has gradually incorporated many technological innovations [5, 6], its overall composition has remained relatively constant and includes the following:

1. launch control,
2. health monitoring, and
3. security control.

NASA aims to further standardize various LCS internal components, with closed-loop real-time control requirements allocated to the front-end and embedded systems. PLCs are widely employed for ground support element and LV operations, PLCs, LVs, and spacecraft, all connect to the back-end control and monitor applications through their gateways. Situational awareness and human-in-the-loop requirements are all allocated to control room servers and workstations.

NASA expects construction of SLS launch control centers to consider three main factors.

1. The launch control system should have 40-year service life.
2. Employ standard equipment, including industrial standard PLCs, not subject to a single supplier.
3. The system should have strong flexibility and adaptability, to cope with future LV and aircraft requirements, which are not currently clear.

6.1.2 Chinese Launch Control System

Remote LCSs incorporating the front-end and back-end configurations have been adopted throughout the Chinese manned spaceflight project, as shown in Fig. 6.1.

Front-end facilities include the following components.

- Test and control cabinets.
- Power supply cabinet.
- Power control assembly: for starting and stopping power, and voltage regulation.
- Launch control assembly: implement control logic with relays.
- Pressure control assembly: for pressurization, and gas pipe disconnection.
- Power supply and distribution assembly: provides power-on, power switch, and power-off controls.
- IMU testing assembly: for stimulation and sampling of inertial meters.
- Data recorder: to automatically record onboard analog signals.
- System interface assembly: to simulate or check external interfaces.
- Ignition unit: output large current for pyrotechnics.
- PLC units and VXI (PXI) assemblies: for logic control instructions and testing.

Power supplies include DC regulated (DCPS) and intermediate frequency inverter power supplies (IFPSs). DCPS include various types depending on the onboard supply demand, with each having dual and parallel redundancy: master and slave units, master voltage slightly higher than the slave; when the master fails, the slave automatically takes overpowering. IFPS operates in a hot–cold configuration, when the hot IFPS fails, the cold one is switched by the power distribution assembly.

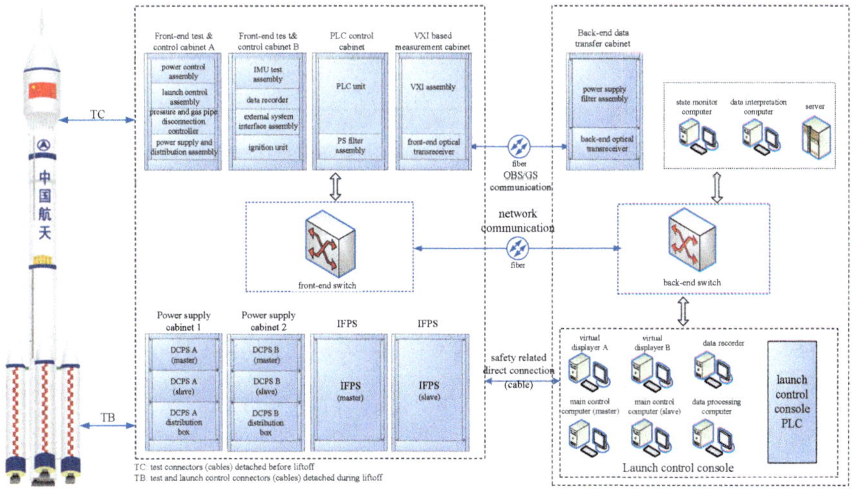

Fig. 6.1 An example of the LV TLCS configuration

To avoid contact loss between the front-end and back-end equipment due to network failure, a direct connection (cable) is retained for safety-related signals and commands in emergency conditions.

6.1.2.1 Launch Control Flow

Launch processes [7] are closely related to the LV design. Although the flow is different for each LV type, the basic design concept is similar. This section considers a "typical" launch process and prelaunch activity, as shown in Fig. 6.2.

- −8 h: Prelaunch processes usually commence 8 h before liftoff (hence "−8 h"). A simplified functional check is performed, which constitutes the first and final overall inspection after fuel filling. After the check, the launch flow passes to the launch state.
- −4 h: Target orbit parameters are determined, and the preferred wind correction trajectory is selected depending on measured upper atmosphere wind speeds. Launch elements are subsequently calculated, incorporating the aiming parameters, to compile the final version FSW. A responsive verification test is conducted for acceptance.
- −2 h: Prelaunch check of the control system, the final and only sampling inspection after the cabin is sealed. If passed, flight software is uploaded to the OBC, checked, and downloaded again for verification on the simulation test platform for the final time.
- −30 min (approx.): FSW starts running self-checking OBC and other equipment and forwarding key details to the TLCS, including launch direction, IMU parameters, index sign related to guidance methods, and software running state.

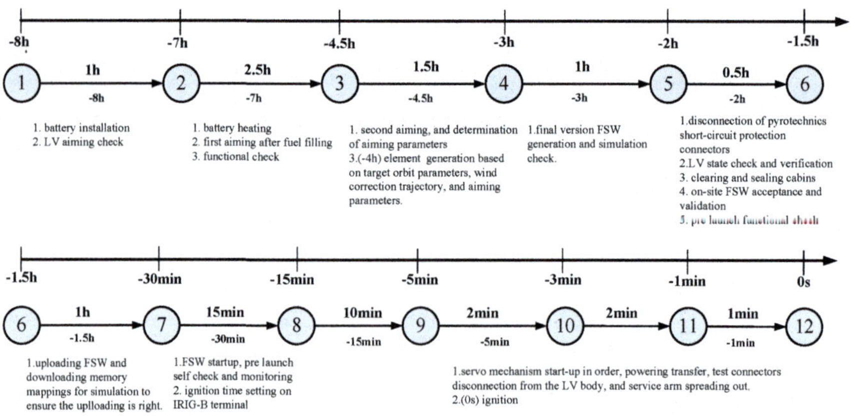

Fig. 6.2 An example of work flow eight hours before launching

If a problem is identified, error encoding is forwarded to GS for fault location and decision-making.

- −15 min: Timing control unit or InterRange Instrumentation Group B code terminal (IRIG-B) is evoked, which is employed for automatic launch control (ignition and emergency shutdown). If the IRIG-B terminal is abnormal, launch control can be switched from automatic to manual mode.
- −5 min: IRIG-B terminal countdown activates the servomechanism, and the power transfer command is issued. Power supply is subsequently switched from ground to onboard battery.
  Once the control console receives the power transfer command, the communication route from GS to OBS is prohibited to avoid interference with FSW. FSW transmits the last self-check results to the TLCS.
  - If the results are abnormal, the launch process is suspended, and the on-site commander decides whether or not to continue the process.
  - If the results are normal, the connectors between GS and OBS are detached, and the service arms spread. The LV is ready for ignition.

  During this period, FSW calculates and sets initial states.
  Permission controls are set up to avoid illegal ignition command, i.e., accepted as TRUE only after a specific time, when the power transfer command is sent. The ignition command is received by either OBS or GS.
- 0 s: The IRIG-B terminal transmits ignition instruction to the front-end PLC, which relays the instruction to the relevant onboard or ground equipment to complete ignition.
  The engine builds thrust, the LV moves away from the launch pad, and take-off contacts are switched on and sampled by the FSW. At this moment, the launch point inertial coordinate system is established and the real-time flight control is activated.
  The LCS automatically transmits emergency shutdown instructions at some fixed period after ignition. Under normal conditions, the LV will have already left the launch tower, and the connectors on the tail of the rocket connecting OBS and GS will have been detached. Hence, the shutdown instructions will not reach the OBS and will have no effect on the flight. However, if the LV has failed to takeoff for any reason, the connectors will remain connected, and the shutdown instructions will reach the OBS, triggering automatic shutdown.

Figure 6.3a, b shows the process from −30 min in further details. Flight control, ground control, and IRIG-B terminal all interact. Symbols ①–⑧ represent fault states that cannot be handled automatically, and it should be considered by the on-site commander. The "A" box shows self-check processes, including compulsory and non-compulsory items. If a fault is identified in a compulsory item, a decision is required immediately, whereas faults in non-compulsory items may be ignored.

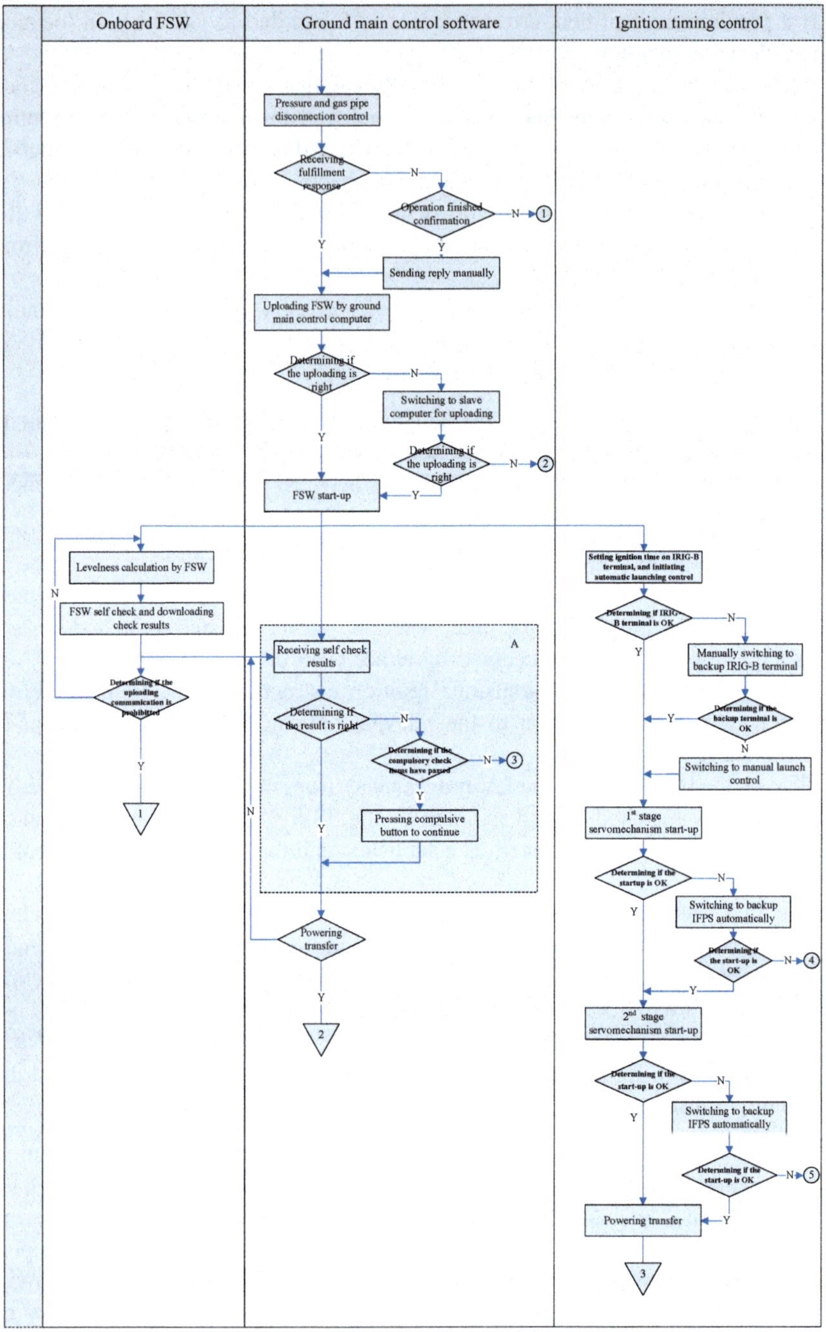

Fig. 6.3 Work flow thirty minutes before launching (**a**). Work flow thirty minutes before launching (**b**)

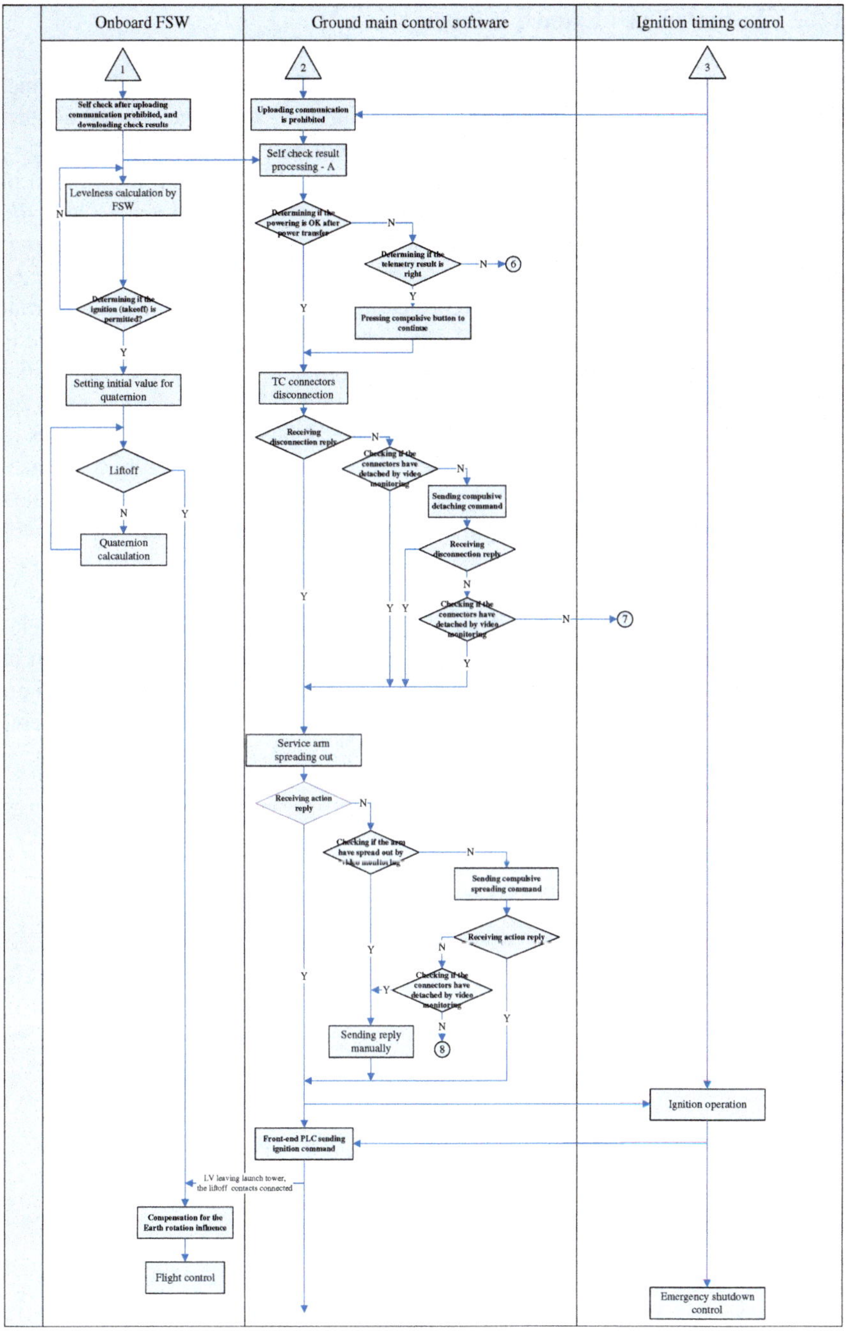

Fig. 6.3 (continued)

6.1.2.2 Instantaneous Launch Window

The instantaneous launch window [8] ensures liftoff on the predesignated time. Since the interval time is relatively fixed from ignition to when the take-off contact is closed, liftoff time is generally converted into ignition time.

The instantaneous launch window is mainly required for high orbit precision in inertial space. If the target orbit relative to inertial space is stationary, since the Earth is spinning, the launch point relative to the target orbit is constantly changing, i.e., the spatial relationship between the LV initial position and target orbit changes. However, flight control requires a reference coordinate, i.e., the launch point inertial coordinate system, which is set up at the moment of takeoff. When launch time is set, this also fixes the launch point inertial coordinate system, and the spatial relationship between LV initial position and target orbit. The launch element is set in accordance with this relationship. If the launch is delayed, the Earth continues to rotate, along with the LV, causing the time-delayed launch point inertial coordinate system to be different from that of the scheduled time. Since the LV is controlled from the new launch point but using the original launch elements, the payload will be sent to a different orbit.

Delayed liftoff affects the right ascension of ascending node (RAAN), and is generally converted to longitude of the ascending node (LAN) in the launch element. The target orbit LAN constantly changes with Earth's rotation, as shown in Fig. 6.4. Let the target orbit LAN for the scheduled liftoff time = $\Omega_{t,s}$, and longitude of the launch point = $\Omega_{l,0}$. Then, the LV will turn $d\Omega = \Omega_{t,s} - \Omega_{l,0}$. If takeoff

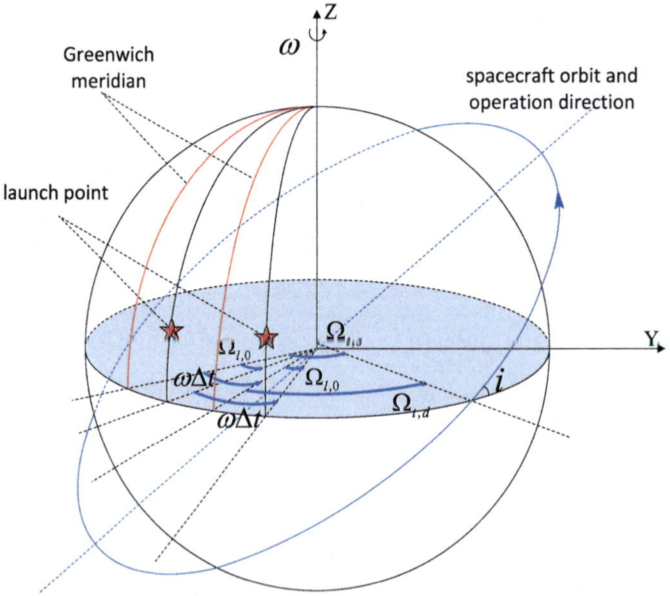

Fig. 6.4 Effect of delayed launch on orbit injection accuracy

is delayed Δt, the Earth, and hence the launch point, rotates $\omega\Delta t$, where ω is the angular rate of the Earth. Then, the LAN of the target orbit is $\Omega_{t,d} = \Omega_{t,s} - \omega\Delta t$, but the launch point longitude remains $\Omega_{l,0}$. If the LV turns $d\Omega$ in accordance with the original launch element, the LAN of the LV orbit injection is

$$\Omega_{l,\text{final}} = \Omega_{l,0} + d\Omega = \Omega_{l,0} + \Omega_{t.s} - \Omega_{l,0} = \Omega_{t.s} = \Omega_{t,d} + \omega\Delta t.$$

Thus, the error between the LAN of the target orbit and the LAN of the LV orbit injection, or orbit injection RAAN error, $e_R = \Omega_{l,\text{final}} - \Omega_{t,d} = \omega\Delta t$, as shown in Fig. 6.4.

Since Earth's rotation rate = 0.004167°/s, if launch is delayed by 1 s, RAAN error is increased by 0.004167°. Thus, if the required orbital accuracy is not high, e.g., allowable error = 0.5°, the effect of a 1-s delay can be almost ignored. On the other hand, if the required accuracy is very high, e.g., allowable error = 0.01°, then a time delay of only 2.5 s will exceed the required threshold.

To make full use of the advantages of the closed-loop guidance method, liftoff time can be accurately determined by GNSS or other device, and the launch elements regulated depending on the difference between actual and theoretical liftoff time. Then, the flight trajectory can be automatically recalculated based on the renewed element, and RAAN error caused by the delay can be eliminated. Thus, the instantaneous launch window can be widened, which is beneficial for arranging the launch, particularly for cryogenic LVs. However, the widened window range depends on the LV's carrying capacity margin.

Precise timing control is provided by IRIG-B terminal to facilitate takeoff (ignition) on schedule. The B-code timing signal provides the time reference for ignition, and the command is converted into discrete signals relayed to the LCS control console.

6.1.2.3 Launch Point Inertial Coordinate System Establishment

The launch point inertial coordinate system [9] can be set depending on the levelness sensed by the accelerometers, which is the angle of the normal or transverse axis relative to the horizontal plane of the Earth, and the initial attitude of the launcher can be expressed in quaternion form as

$$\varphi_0 = \frac{\pi}{2} + \Delta\varphi_0, \tag{6.1}$$

$$\begin{cases} q_0 = \cos\frac{\gamma_0}{2}\cdot\cos\frac{\psi_0}{2}\cdot\cos\frac{\varphi_0}{2} + \sin\frac{\gamma_0}{2}\cdot\sin\frac{\psi_0}{2}\cdot\sin\frac{\varphi_0}{2} \\ q_1 = \sin\frac{\gamma_0}{2}\cdot\cos\frac{\psi_0}{2}\cdot\cos\frac{\varphi_0}{2} - cos\frac{\gamma_0}{2}\cdot\sin\frac{\psi_0}{2}\cdot\sin\frac{\varphi_0}{2} \\ q_2 = \cos\frac{\gamma_0}{2}\cdot\sin\frac{\psi_0}{2}\cdot\cos\frac{\varphi_0}{2} + \sin\frac{\gamma_0}{2}\cdot\cos\frac{\psi_0}{2}\cdot\sin\frac{\varphi_0}{2} \\ q_3 = \cos\frac{\gamma_0}{2}\cdot\cos\frac{\psi_0}{2}\cdot\sin\frac{\varphi_0}{2} - \sin\frac{\gamma_0}{2}\cdot\sin\frac{\psi_0}{2}\cdot\cos\frac{\varphi_0}{2} \end{cases} \tag{6.2}$$

where $\Delta\varphi_0$, and ψ_0 are levelness measurement (see Eqs. 4.20 and 4.21), respectively; γ_0 is initial roll angle, $\gamma_0 = A_0 - A_l$.

The transformation matrix of the LV coordinate system to the launch point inertial coordinate system, $\boldsymbol{A}$, can be expressed as

$$\mathbf{A} = \begin{bmatrix} A_{11} & A_{12} & A_{13} \\ A_{21} & A_{22} & A_{23} \\ A_{31} & A_{32} & A_{33} \end{bmatrix} \\ = \begin{bmatrix} q_0^2+q_1^2-q_2^2-q_3^2 & 2(q_1q_2-q_0q_3) & 2(q_0q_2+q_1q_3) \\ 2(q_1q_2+q_0q_3) & q_0^2-q_1^2+q_2^2-q_3^2 & 2(q_2q_3-q_0q_1) \\ 2(q_1q_3-q_0q_2) & 2(q_0q_1+q_2q_3) & q_0^2-q_1^2-q_2^2+q_3^2 \end{bmatrix} \tag{6.3}$$

In the launch point inertial coordinate system, the navigation initial value (initial velocity and position) can be determined by the launch point coordinates (A_0, B_0, H_0). Since the coordinate origin is at the launch point, the initial position is zero, but speed is nonzero due to Earth's rotation in the inertial coordinate system. The geocentric radius vector and initial velocity of the origin can be expressed as

$$\begin{cases} V_{x0} = \omega_E\left[H_0 + R_a\left(1+\alpha_E \sin^2 B_0\right)\right]\cos B_0 \sin A_0 \\ V_{y0} = 0 \\ V_{z0} = \omega_E\left[H_0 + R_a\left(1+\alpha_E \sin^2 B_0\right)\right]\cos B_0 \cos A_0 \\ R_0 = \left\{R_a(1-\alpha_E)/[1-\alpha_E((2-\alpha_E)]\cos^2\varphi_{dx0}]^{1/2}\right\} + H_0 \\ \varphi_{dx0} = tg^{-1}\left[(1-\alpha_E'^2)tgB_0\right] \\ \mu = B_0 - \varphi_{dx0} \\ \alpha_E'^2 = \alpha_E(2-\alpha_E) \\ R_{0x} = -R_0 \sin\mu \cos A_0 \\ R_{0y} = R_0 \cos\mu \\ R_{0z} = R_0 \sin\mu \sin A_0 \end{cases} \tag{6.4}$$

where V_{x0}, V_{y0}, V_{z0} represents the velocity components of the origin in the launch point inertial coordinate system, respectively; usually, the origin is located where the IMU is installed. ω_E is the Earth's rotation rate, α_E is the Earth's oblateness, H_0 is the height from the launch inertial coordinate system origin to the ground (launch site), R_0 is the distance from the origin to the Earth's center, R_{0x}, R_{0y}, R_{0z} represent the components of R_0 in the launch point inertial coordinate system, respectively. The definition of A_0, B_0, R_a, and φ_{dx0} can refer to Eqs. (4.22)–(4.23).

6.2 Front-End Launch Control System

The LCS is configured with front-end and back-end facilities for safety reasons, which was first adopted in China by the Chinese manned spaceflight project. The front-end facilities are mainly testing, launch control, and power cabinets. Many of the assembled units and modules installed in the cabinets are actuators for control center commands, such as launch control, power-on and distribution, ignition, and emergency shutdown control.

6.2.1 Logic Control and Instruction Circuit Design

The logic control and instruction circuit is the basic unit for launch control. Each command must meet certain preconditions to function, otherwise, even the command is sent, it will not produce valid output. These preconditions, or chain conditions are realized by the logic circuit.

The logic control and instruction circuit usually comprise relays and PLCs. Before the 1960s, relays were more commonly used. Relay systems are simple and intuitive, and can achieve greater power conversion, but are inconvenient for design changes, and do not allow rapid restructuring on-site and other requirements. Therefore, PLC technology was developed for industrial control and gradually replaced relay systems. Modern LCS incorporated relays and PLCs [10–12], with relays particularly applied for high-power output control.

PLCs are designed for digital operation and mainly employed for discrete control, logic control, communication, and signal sampling. The most common PLC programming language is the ladder diagram, and ladder diagram processing methods include hardware logic solvers (HLS) and soft problem solving, also known as compiling and explanatory methods, respectively. The ladder diagram itself is a bit logic operation, and HLS uses a dedicated core rather than the general CPU to process ladder diagram programs, while soft solutions use a general-purpose processor. However, general-purpose processors are inefficient for bit operation, and processor performance must be increased to compensate for this. HLS systems commonly employ dual-core processors [13], one being used as a general-purpose processor for control functions, responsible for system startup, main operation, state query and transfer, function block operation, interrupt processing, and refreshing the IO state table; while the other is dedicated to solve ladder diagrams.

Current commercial PLCs commonly adopt soft solutions with high-performance processors. However, this leads to larger physical size and higher power consumption, and the increased heat dissipation is problematic for sealed design. In addition, some technical details of commercial products, particularly redundant management are not openly accessible, which is unsuitable for aerospace applications, where high reliability and safety are absolute concerns.

In contrast, HLS has advantages of low power consumption, small physical size, and sealed design, making it preferable for hostile environments. Generally, HLS advantages include the following aspects.

1. Dedicated instructions for faster resolving.
   A dedicated instruction set is designed for bit operation. For example, the AND operation requires a single instruction under HLS, whereas the following steps are required by software-based explanatory method, with consequentially less efficiency and speed.

   i. Fetch instruction.
   ii. Determine the instruction type.
   iii. Calculate the absolute address of the word, where the component state is stored based on the word offset address.
   iv. Obtain the value of the word depending on the absolute address.
   v. Determine the current status of the component from one bit of the word based on the bit offset address.
   vi. Run the AND operation with the current status of the component (on or off) and the input.
   vii. Write the AND operation result to the associated register.

2. Dedicated register window.
   Various registers are defined depending on ladder diagram solving requirements, such as the column status register, which is designed in the register window form, where the output of the previous column is the input to the next column, and there is no need to repeatedly access the memory.
3. Parallel operation.
   Vertical scanning of the ladder diagram is suitable for parallel operation, with a dedicated hardware circuit typically employed for parallel operation.
4. Reliability design.
   HLSs are not subject to the general processor architecture, and targeted reliability design can be adopted to meet high-reliability requirements.

The remainder of this section focuses on HLS-based PLC design and applications.

6.2.1.1 Ladder Diagram Solving

There are two methods to solve the ladder diagram. Horizontal scanning is suitable for Boolean algebra and instruction list logic, but compiler requirements are very high, some network aspects cannot be supported, and there are significant user restrictions. On the other hand, vertical scanning, as shown in Fig. 6.5, allows AND and OR operations on the same column to be performed in one instruction period, which is equivalent to parallel execution in the same network. This scanning mode follows the principle of physical electrical diagrams and is most suitable for diagram solving. It also has very low ladder diagram composition (programming) requirements and can solve almost any ladder diagram.

The ladder diagram is converted into instruction or machine code by the compiling software, and the compiling process is directed at a specific encoding rule.

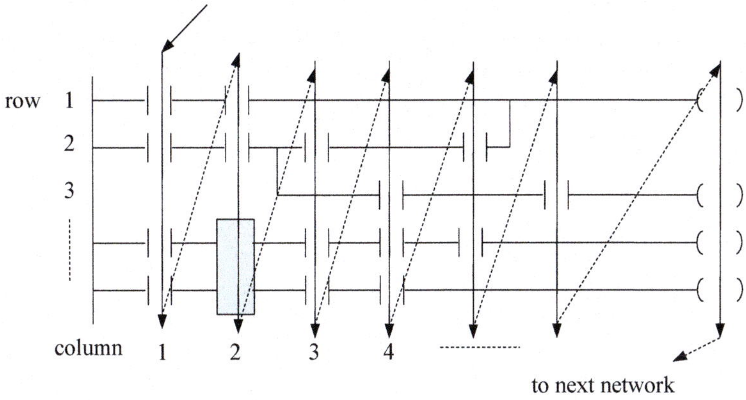

Fig. 6.5 Vertical scanning mode

D15	D14	D13	D12	D11	D10	D9	D8	D7	D6	D5	D4	D3	D2	D1	D0	Description
0	0	0	Word offset address 0-512									bit offset address				-| |-
0	0	1														-|N|-
			......													
1	0	0														-( )-
1	0	1	V5	V4	V3	V2	V1	V0	E6	E5	E4	E3	E2	E1	E0	BOC
			OR relationship						component state (if there's a component in the row)							
			......													
1	1	1	1	Label #0~4095												SON & Label
				Functional block												
				......												

Fig. 6.6 An example of ladder diagram instruction encoding

HLS encoding rules differ for different HLS designs. Most users do not care about the encoding rules, and design tools automatically convert the diagram into the final code.

Figure 6.6 shows an example instruction encoding rule to illustrate the HLS principle. In this instruction system, D15–D13 indicate the type of instruction, where 000 indicates the component element is a normal open node, i.e., -| |-, and D12–D4 is the status word offset address of the component. The word value can be read based on this offset adding the base address of the word. Each status word has 16 bits, which can represent the status of 16 components. D3–D0 indicate the position of the current component in the word.

Consider the example shown in Fig. 6.7. While a complete HLS instruction set should also include Constant, Register, Pointer, Function, Extension, etc., only

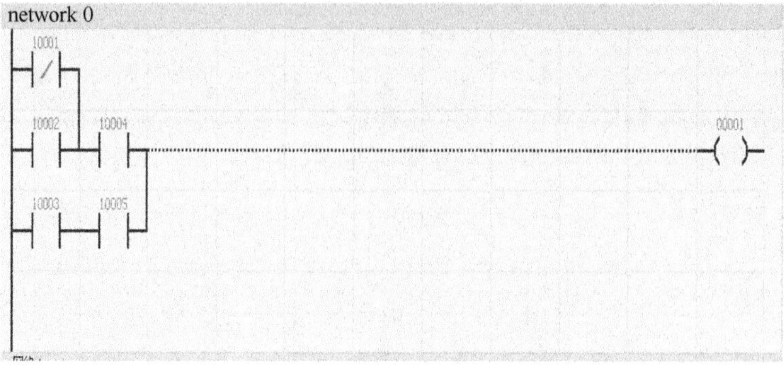

Fig. 6.7 A simple example of a ladder diagram program

those used in the example are discussed here. The ladder diagram consists of three columns, and the compiler forms the instruction code after vertical scanning, as shown in Table 6.1.

HLS executes the ladder program sequentially based on Table 6.1, where the process includes analysis, execution, and write back. The example is a good illustration of parallel column operation in vertical scanning mode. The ladder instructions are summarized as follows.

1. Instruction 1: the SON instruction indicates the start of a ladder diagram page. HLS initializes the first column component input to 1.
2. Instruction 2: the BOC command indicates the start of the first column. Following the definition in Fig. 6.6, this instruction indicates there are components in the first three rows, and the components of the first and second row have an "OR" relationship.
3. Instructions 3–5: node instructions to calculate outputs depending on external nodes 10001–10003 (relay, switch, etc.) states and the input state (already initialized to 1). After the nodes are executed, HLS calculates all output states of the column depending on the "OR" relationship in one clock cycle.
4. Instruction 6: the start of the second column, showing there are components on the second and the third row that have an "OR" relationship.
5. Instructions 7, 8: node instructions to calculate the outputs of nodes 10004, 10005 depending on their states (relay, switch, etc.) and the input state (i.e., the output state nodes 10002, 10003). When the nodes are executed, HLS calculates all the output states of the column depending on this column components "OR" relationship in one clock cycle.
6. Instruction 9: the start of the third column with only one coil, 00001, whose state is determined by the input (i.e., the output of second column and second-row component, 10004).

Table 6.1 Ladder diagram program (Fig. 6.7) parsing

Serial number	Instruction	Instruction code	Description
1	SON	F000H (1111 0000 0000 0000B)	Network start-up code, the first network (numbered from 0)
2	BOC 111 0000	A087H (1010 0000 1000 0111B)	The first column start-up code, indicating there are components in the first three rows, and the components of the first and second row have an "OR" relationship
3	-|\\\|- 10001	2000H (0010 0000 0000 0000B)	Class B node, fetching the state of node 10001
4	-| |- 10002	0001H (0000 0000 0000 0001B)	Class A node, fetching the state of node 10002
5	-| |- 10003	0002H (0000 0000 0000 0010B)	Class A node, fetching the state of node 10003
6	BOC 0110000	A106H (1010 0001 0000 0110B)	The second column start-up code, indicating there are components on the second and the third row, and the components have an "OR" relationship
7	-| |- 10004	0003H (0000 0000 0000 0011B)	Class A node, fetching the state of node 10004
8	-| |- 10005	0004H (0000 0000 0000 0100B)	Class A node, fetching the state of node 10005
9	BOC 010 0000	A002H (1010 0000 0000 0010B)	The third column start-up code, indicating there is component on the second row
10	() 00001	8800H (1000 1000 0000 0000B)	Node 00001 state output, the node's address is 2049
…	…	…	…

6.2.1.2 PLC Applications in LCS

Due to their flexibility, PLC functions in LCS can be divided into the following categories.

1. Control command outputs, such as power on and off, power distribution, power switch, equipment start and stop, etc.
2. Sequential command outputs, i.e., combination of control and timing commands, such as ignition, shutdown (emergency shutdown), etc.

3. Conditional control functions used to set up various states, which are often regarded as chain conditions in process control.
4. State sampling and indication, and command response.
5. Analog signal sampling and output.
6. Direct communication between front-end and back-end PLCs.

Redundancy design is commonly adopted to ensure launch control reliability. PLCs use modules and backplane configurations, where a chassis is not required, and typical configurations include a backplane (motherboard), embedded controller (CPU module), discrete input and output (DI/DO) module, analog signal module (with conditioning and isolation circuits), and DC power supply module.

The following sections introduce PLC configuration, redundancy design, and operation modes.

PLC Configuration

Front-end and back-end PLC configurations can be set in the following modes.

1. Applications (ladder diagram program) are installed on both ends.
   Front-end PLCs receive OBC feedback, and commands from back-end PLCs, then run logic operations and transmit the calculated results and corresponding state(s) to back-end PLCs. Back-end PLCs receive staff or operator inputs and information from front-end PLCs, run logic operations, and transmit the outputs to display terminals and front-end PLCs. Application software running on both PLCs and computation is localized.
2. Front-end PLCs only perform I/O mapping for back-end PLCs.
   Front-end environment requirements are significantly more rigorous than back-end, hence front-end PLC assemblies should be simplified, and do not run ladder diagrams; only their MAC and IP address are configured. All programs run on back-end PLCs, but front-end and back-end I/O modules are mutually mapped. Thus, front-end inputs can be sampled synchronously at the back-end, and back-end generated output is synchronously mapped to front-end I/O modules for output control.

In mode 1, front-end and back-end PLCs have stronger operational ability and flexibility, but front-end PLCs are more complicated. In mode 2, front-end PLCs are simplified, and I/O mapping function is accomplished by a general-purpose processor without requiring HLS. However, all logical operations are run on the back-end, hence communication between the front-end and back-end is frequent.

PLC Redundant Design

PLC failure is generally open circuit (i.e., no output) or short circuit (i.e., wrong output). If the failure probabilities of the two modes and harm level are the same, a TMR voting scheme is more reasonable, particularly for some key signals.

If the probability of short-circuit fault is larger, series connection is preferable, whereas if the probability of open circuit fault is larger, parallel design is adopted. To further refine reliability design, comprehensive series-parallel or parallel-series design is adopted, depending on the harm level. Dual redundancy can simplify reliability design, but it is based on the assumption that certain failure modes are very rare or the harm result is endurable. Under this condition, some chain conditions and status monitoring are required.

Master–Slave and Parallel Operation Modes

Dual redundant design includes master–slave and parallel operation. For master–slave operation, one of the two CPU modules produces real output, which is self-checked and checked by the other through network communication. If the main CPU module fails, the system automatically switches to the slave module. Since diagnosis is involved, redundant management is complex.

The parallel work mode is intuitive and simple. Consider the example shown in Fig. 6.8. The back-end PLC works in parallel mode, with instructions from the main control computer for automatic control, ignition instruction from the IRIG-B terminal, and instructions sent manually through control console buttons, are both relayed to two back-end PLCs, and then mapped to the front-end PLCs. The front-end PLCs construct a parallel-series connection to output various commands, as shown in Fig. 6.9, where PLC1 and PLC2 refer to front-end PLCs, and K1–K4 refer to the corresponding output units.

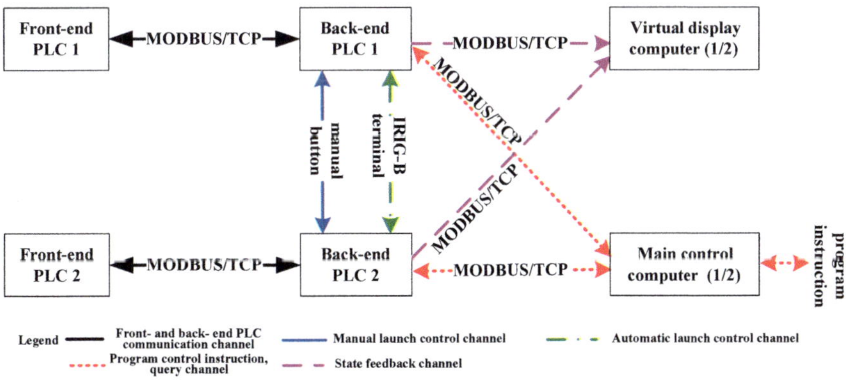

Fig. 6.8 Dual PLC parallel working principle diagram

Fig. 6.9 PLC parallel-series output schematic diagram

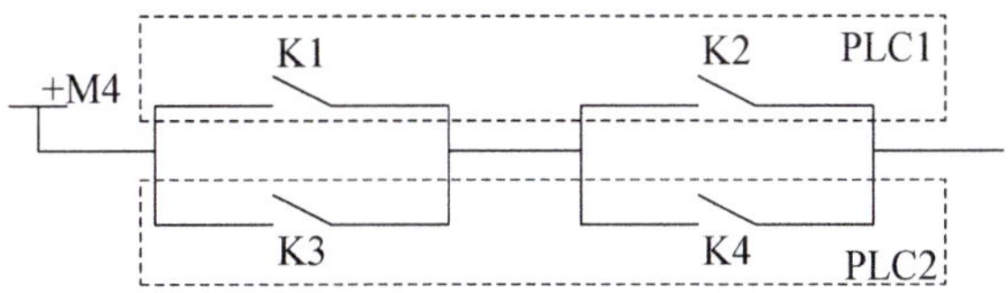

Any I/O unit failures of the front-end and back-end PLCs will not affect system function. However, if all units are 0 or 1 for some reason (e.g., due to CPU fault), the parallel-series design will not work, and different chain conditions must be used for this failure mode.

The PLC operational mode should be based on the application situation, and failure modes and consequences. Although PLCs are more convenient than relay devices, this comes at some cost, such as software development and verification, which should be considered.

6.2.2 Power Supply and Distribution Design

For most ground tests and prelaunch preparations, the OBS is powered by ground power supplies, including powering for onboard avionics and pyrotechnics/valves. Thus, the ground power system is set depending on the OBS configuration, with the variety of PSs minimized. This section discusses the powering for avinoics, which can be achieved in direct or feedback mode.

6.2.2.1 Direct Ground Powering Mode

The direct mode is used where OBS current requirements are relatively small, and the exclusive power cables exist. The voltage regulating point is very important for the PS system, and the point where stable voltage is maintained by closed-loop control is set onboard (usually on the PDU terminals). If OBS current consumption is small, the voltage drop consumed on the powering cables from GS to OBS is also small. On the other hand, the powering cables connected directly to an outer bulkhead plug near the onboard power consumption equipment is designed thicker to lower resistance and voltage drop. The cables and plug are disconnected after power switching from GS to OBS, without increasing OBS weight and affecting liftoff. Both of the designs ensure the ground PS voltage is not very high on the local end, and can also power the TLCS itself.

For the power switch simulation test, chemical battery life cycles are limited and could not participate in testing every time, so another dedicated power cable connected directly to the ground PS may replace the battery, referred to as the simulated power cable. The voltage supplied by the cable is slightly higher than that of the onboard regulating point to meet power switch simulation test requirements.

However, this design is not suitable for all applications. For example, if outer bulkhead plugs are removed to simplify operations, and only cable plugs are retained at the LV tail, then the power cable must be installed onboard from the LV tail to the instrument cabin, which raises the following problems.

1. Voltage drop is large, requiring high voltage on the ground PS local end.
   The power cable from the LV tail to the top is longer and has large resistance. If OBS current is large, the voltage drop along the cable will be large. Hence, to ensure OBS equipment remains at 28 V, the ground power supply voltage must be higher, even up to 40 V. The ground PS usually also powers TLCS devices that work under 28 V, so this will lead to ground devices working at higher voltage, or exceeding component upper limits; both are unfavorable for TLCS reliability.
2. Power cables are heavy, which affects LV carrying capacity, and it is inconvenient for cables to be laid along the LV outer surface.
   To reduce voltage drop as much as possible, the powering cable should be thickened to reduce impedance, which increases the cable weight. In some situations, the cable is partially laid along the LV outer surface, which will require heat shielding, and thicker cable requires larger shielding. This not only increases unreliable factors for shield structural installation but also affects LV aerodynamic performance.
3. Voltage changes greatly for different operational and load states, making it difficult to meet all requirements.
   Due to the cable resistance, when OBS current consumption changes in the powering-on process, particularly the IMU heating current varies greatly between initial and steady states, local PS voltage can fluctuate greatly, and ground facility power stability cannot be guaranteed.
4. Transient high voltage will impact the OBS during power switch with simulated power cable.
   As discussed above, the simulated power cable mainly replaces the onboard battery for power supply switch during system level tests. When switching, the voltage regulating point remains onboard and is not instantaneously switched. Hence, the cable provides the local end voltage, which is much higher than 28 V, producing transient high voltage impacts. Only when the voltage regulating point switches to the local PS end can the cable voltage drop to 28 V.

Given these issues, onboard feedback is generally adopted.

6.2.2.2 Onboard Powering Feedback Mode

As above, the ground power supply is adopted and the voltage regulating point is set onboard to ensure OBS voltage stability. However, GS devices are not directly powered by the ground PS, and a powering bus is introduced or fedback from the OBS to the ground facilities. Thus, even if the local PS voltage is very high, it will not affect the ground facilities. The underlying assumption is that the OBS power load greatly exceeds the GS.

Figure 6.10 shows the powering schematic before power switch, where TA + and TA- are ground PS connectors (the PS is denoted as M1), and the voltage regulating point is set on the PDU input terminal, which ensures stable voltage

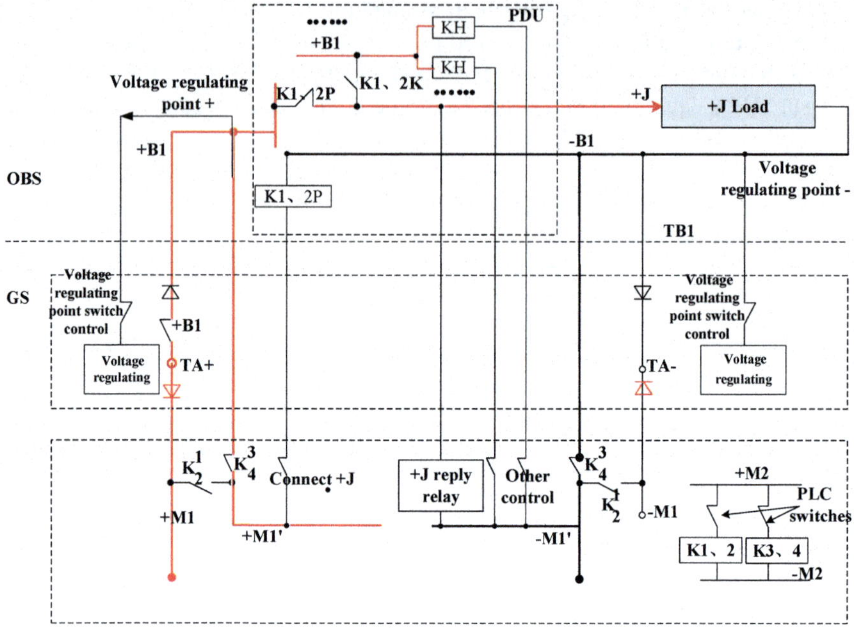

Fig. 6.10 Powering and control schematic before power switch

onboard under load changes. The stable onboard PS bus is fedback (+M1') to service ground facilities.

Only those ground facilities that require the same powering bus as the OBS are powered from the feedback bus, other equipment can use an independent ground PS (s). The ground power load is small enough to allow a thinner feedback bus cable while retaining acceptable voltage drop. The upgoing power cable from the LV tail to onboard avionics is also greatly simplified because the onboard regulating point ensures stable voltage for the OBS, regardless of cable resistance and the voltage on the ground PS ends. Even if the thinner upgoing power cable requires higher PS voltage, the voltage no longer directly impacts the ground facilities. Thus, OBS and GS powering requirements are met while reducing onboard cable weight.

Figure 6.11 shows the powering schematic after power switch. When the switch command is sent, the OBS is powered on by the onboard battery (or the simulated power cable). There is a period when the ground PS and battery are in parallel, with the one with higher voltage producing real load output. However, this load causes voltage drop, and then the other also produces real output. This process alternates until the two power sources balance and they both provide load current. When the switch is confirmed, the voltage regulating point is cut back to the local end of the ground PS, i.e., the voltage on the +M1 bus is restored to 28 V. Relay contacts K1 and K2 are subsequently closed, and the ground equipment is powered by the M1 bus and the onboard battery in parallel. Then contacts K3 and K4 are disconnected,

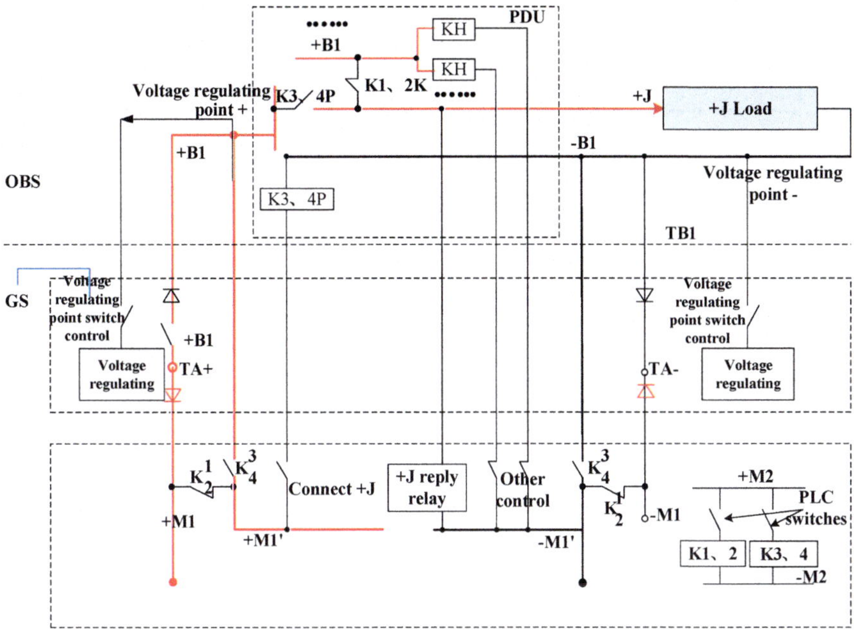

Fig. 6.11 Powering and control schematic after power switch

and only the M1 bus, which supplies 28 V within the ground equipment nominal range, is provided for the GS.

The feedback design focuses on the situation that OBS current consumption is high, and cables and plugs between the OBS and GS only remain on the tail of LV, not on the instrument cabin outer bulkhead, and significantly reduces onboard cable weight.

6.2.3 *Ignition and Emergency Shutdown Control*

Ignition and shutdown control differs for different engines. If the engine is triggered by an electric explosion valve, the valve's on-off state is usually controlled by the detonating of the pyrotechnics, and the detonation is singular and irreversible; if by electromagnetic valve, the state transfer is achieved by the power on or off to the valve, the state is maintained and convertible. Therefore, there are two modes for the ignition and emergency shutdown control.

6.2.3.1 Ground System Instigation

Traditionally, engine ignition is instigated by GS [14]. The pyrotechnics are detonated and power is maintained for a very short period. Figure 6.12 shows a typical

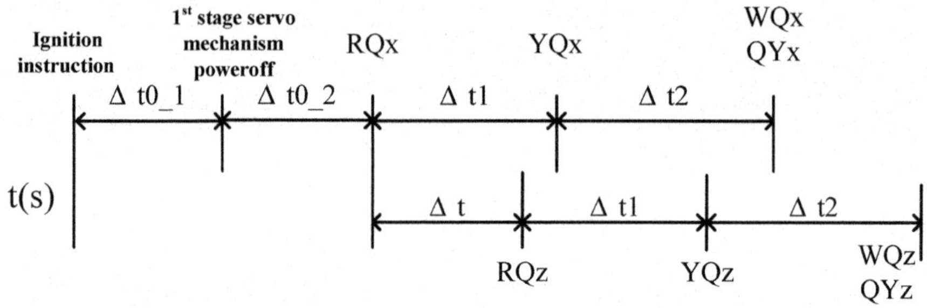

Fig. 6.12 Typical ignition process

example. Once engine ignition is triggered, LV first-stage servomechanisms are powered off after Δt0_1, the core-stage engine starts after Δt0_2, and the booster engine starts after Δt. Core-stage and booster engine starting sequences are the same, first, the fuel valve (RQ) opens, then, the oxidizer valve (YQ) opens after Δt1, and the turbopump valve (WQ) and the reinforcing pressure valve (QY) open after Δt2.

The IRIG-B terminal generates the ignition command. When the countdown ends, the terminal transmits a discrete signal to the back-end PLC, the PLC generates the command sequence and forwards this to the front-end PLC through the network. The front-end PLC drives and controls onboard high-power relays to output the required current to fire the pyrotechnics, using onboard batteries since power switching was previously completed. All sequential instructions are conducted by the GS.

Figure 6.13 shows a typical emergency shutdown process. When the last sequential instruction of the ignition process occurs, an emergency shutdown instruction is automatically sent by the LCS after Δt'. If the LV fails to takeoff from the launch tower in time, i.e., the last plugs or OBS-GS connection cable on the LV tail were not pulled, the shutdown command reaches the OBS through the cable.

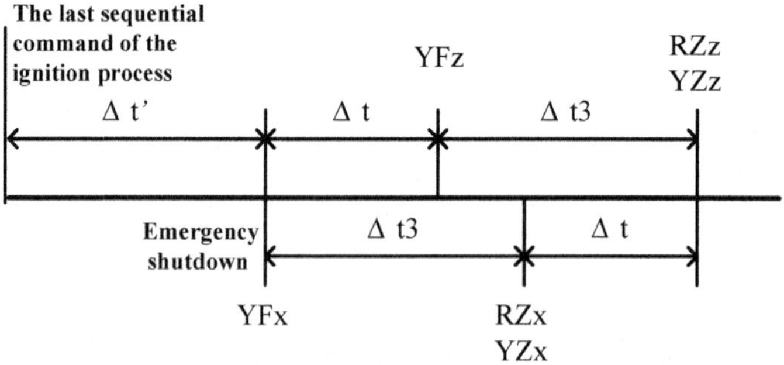

Fig. 6.13 Typical emergency shutdown process

The core engine shuts down first, and the booster engine shuts down after Δt. The shutdown sequence is the same in both engines: the oxidizer subsystem (YF) shuts down first, then the oxidant master system (YZ), and the fuel master system (RZ) shut down after Δt3.

Since the problem that the LV didn't takeoff may be caused by onboard control system, the emergency shutdown cannot totally rely on the onboard devices, which means the emergency shutdown procedure is very different from the ignition procedure. A dedicated ground PS is installed in the front-end, and the emergency shutdown current is drawn from this, through ground high-power relays (rather than onboard relays) and the connection cable, to the onboard pyrotechnics. The shutdown sequences and relays are controlled by a front-end PLC.

Under normal circumstances, the plugs (and cable) disconnect when the emergency shutdown command is sent, and the signal will not have any influence on launcher takeoff.

6.2.3.2 Combined Instigation

Cryogenic LVs generally control ignition and shutdown through power on or off to electromagnetic (or electro-hydraulic) valves. If control is completed by GS, various valve states must be maintained by power switch during connection separation between the OBS and GS, which significantly complicates the design. Moreover, allowing OBS to control ignition will simplify the OBS and GS interface.

The process command is initiated from GS and sent to onboard ICUs. Then, the ICUs implement the ignition process by sequential control. Figure 6.14 shows a typical example of a cryogenic engine ignition process, where the up arrow means power-on, and down arrow indicates power-off. The process is as follows.

−t1 (t1 s before ignition start): power-on to turn off valve DY (electro-hydraulic valve).
−t2: Power-on to open valve DQ1 (fuel drain).
−t3: Power-on to open valve DQ6 (nitrogen flow).

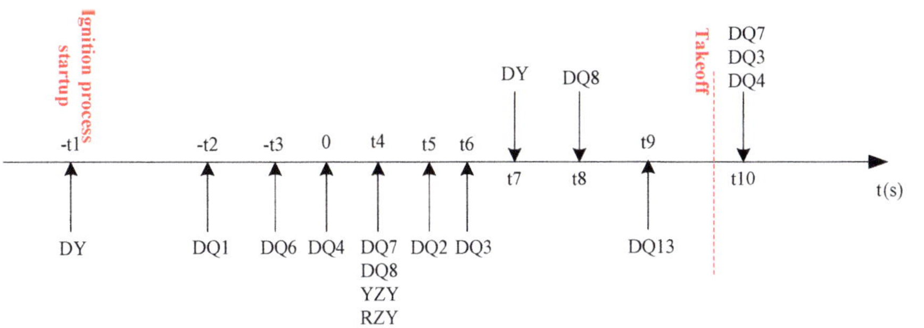

Fig. 6.14 Onboard cryogenic engine ignition process

0 (start ignition): Power-on to open valve DQ4 (to make the control gas to squeeze the starter chassis). Igniter fills the ignition pipeline of the thrust chamber and the front pipeline of the generator fuel valve.
t4: Power-on to open valve DQ7 (liquid oxygen); power-on to open valve DQ8 (fuel throttling in large flow resistance state); power-on pressurization electromagnetic valves YZY and RZY (fuel and oxidizer tanks).
t5: Power-on to open valve DQ2 (generator fuel); Igniter enters the generator; the entry of ignition pipeline of the thrust chamber switches from the entry of generator fuel valve to the entrance of the flow regulator.
t6: Power-on to open valve DQ3 (main thrust chamber fuel valve).
t7: Power-off to close valve DY, flow regulator starts transition.
t8: Power-off to close valve DQ8, fuel throttling changes to a small-flow resistance state.
t9: Power-on to open valve DQ13 (gas blowing), blows the pump isolation chamber using the engine's own gas sources, and the launcher should takeoff normally.
t10: Power-off to close valves DQ7, DQ3, and DQ4, liquid oxygen main valve and the main fuel valve of the thrust chamber are opened under hydraulic pressure.

Emergency shutdown can be achieved only when the various valves are powered off or on, as appropriate. If valve states are maintained by OBS, then GS cannot directly alter them. Therefore, emergency shutdown can only be conducted by the OBS. Thus, if the failure occurred in the sequential control subsystem, e.g., the ICUs, emergency shutdown will fail. The only remaining solution is immediate power-off for the whole system, which may affect engine reuse.

6.3 Back-End Launch Control System

The back-end LCS comprises the control console, a variety of command control and information display terminals, IRIG-B terminal, etc., where manual or automatic launch control and testing is conducted.

The command control device executes various operator instructions, receives automatic test and launch control instructions from the main control computer, and transmits various instructions through the network to the front-end actuators. The command control device is composed of relays and/or PLCs.

The display terminal indicates the current working state of the command control device, i.e., whether or not instructions have been executed, as indicated by feedback sampling. These states, along with other data, are displayed.

6.3.1 Launch Control Console

Operation panels of the launch control console are mainly buttons and switches for manual operation. Most manual operations can also be automatically controlled, but manual operation provides greater flexibility. These operations generally include the following.

1. Remote control of ground PS.
2. Power on and off to ground facilities.
3. Power-on, power switch, and power-off to onboard avionics.
4. Ignition, emergency shutdown, urgent control, etc.
5. Reset control for onboard and ground equipment.
6. Heating control for batteries and inertial equipment.
7. Redundant switch, including ground PS, communication, main control computer, IRIG-B terminal, etc.
8. Pressurization control, connectors, and cables detaching control, etc.
9. State alternation control for various test and launch status.
10. Stimuli for test requirements, such as fuel exhaust simulation, constant current excitation, etc.

6.3.2 IRIG-B Terminal

The IRIG-B terminal provides accurate timing to ensure takeoff (ignition) occurs on time [15]. The terminal uses the IRIG-B timing signal as the ignition control reference, and the B-code timing signal is converted into discrete control signals for the control console. The terminal is usually placed in the back-end, and the time is set by the operator before firing. It can also be installed at the front-end, and the time set through the network without direct personnel intervention. Figure 6.15 shows a typical IRIG-B terminal functional block diagram.

The terminal time base (IRIG-B time) [16, 17] is obtained by decoding the external DC or AC time sources. The DC source is the pulse signal, and the AC source is the modulation signal. If the B code signal is not rcccivcd at some point, the terminal continues timing based on the local clock, referred to as time-keeping, and flags the source missing warning, which recovers when the signal is restored. DC source encoding is shown in Fig. 6.16.

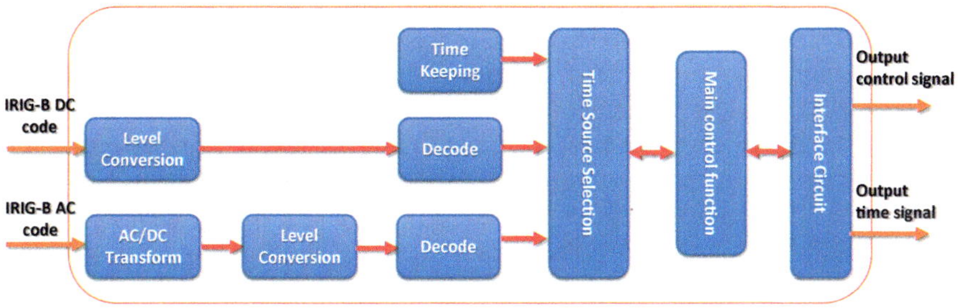

Fig. 6.15 IRIG-B terminal block diagram

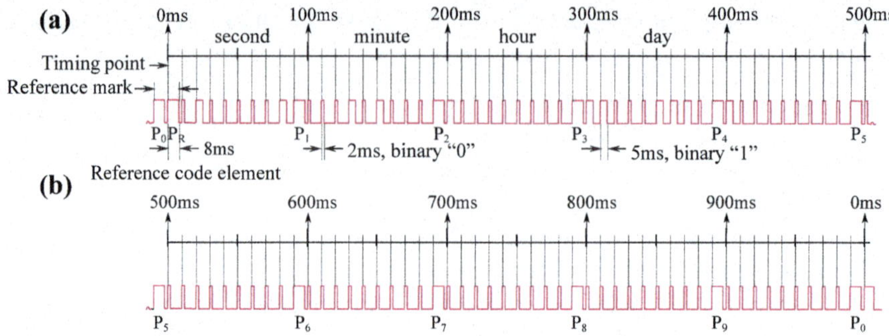

Fig. 6.16 DC source encoding

The focus of terminal design is decoding. For convenience, only DC source decoding is discussed here. In practice, the AC source is first transformed to a DC signal, and subsequent processes follow normal DC decoding. The B-code is updated every second, including the following three special markers in a continuous pulse sequence.

1. Reference code element: 8 ms pulse width.
2. Reference mark: continuous two reference code element.
3. Timing point: rising edge of the second pulse in the reference mark.
4. 2 ms pulse width means binary 0.
5. 5 ms pulse width means binary 1.

Figure 6.17 shows the DC process, which can be summarized as follows.

- Find the reference code element: pulse width at the falling edge of the pulse = 8 ms.
- Find the reference mark related to the reference code element: time interval from the falling edge of this reference element to the previous reference code element = 10 ms.
- Given the reference mark, identify the timing point.
- Decode subsequent pulses.

However, decoding an entire day-hour-minute-second string requires 400 ms, which delays the timing trigger. Therefore, once the timing point is verified, the terminal increases the latest verified time (LVT) by 1 s, and transmits that result. After decoding (approximately 0.4 s elapsed), the terminal examines if the former treatment is correct, and sets current time = LVT. If the timing point is not verified or missing for a period, the terminal continues timing based on its internal clock and LVT until the timing point is verified again and LVT is refreshed. Thus, the combination of external timing sources and internal clock means timing remains uninterrupted.

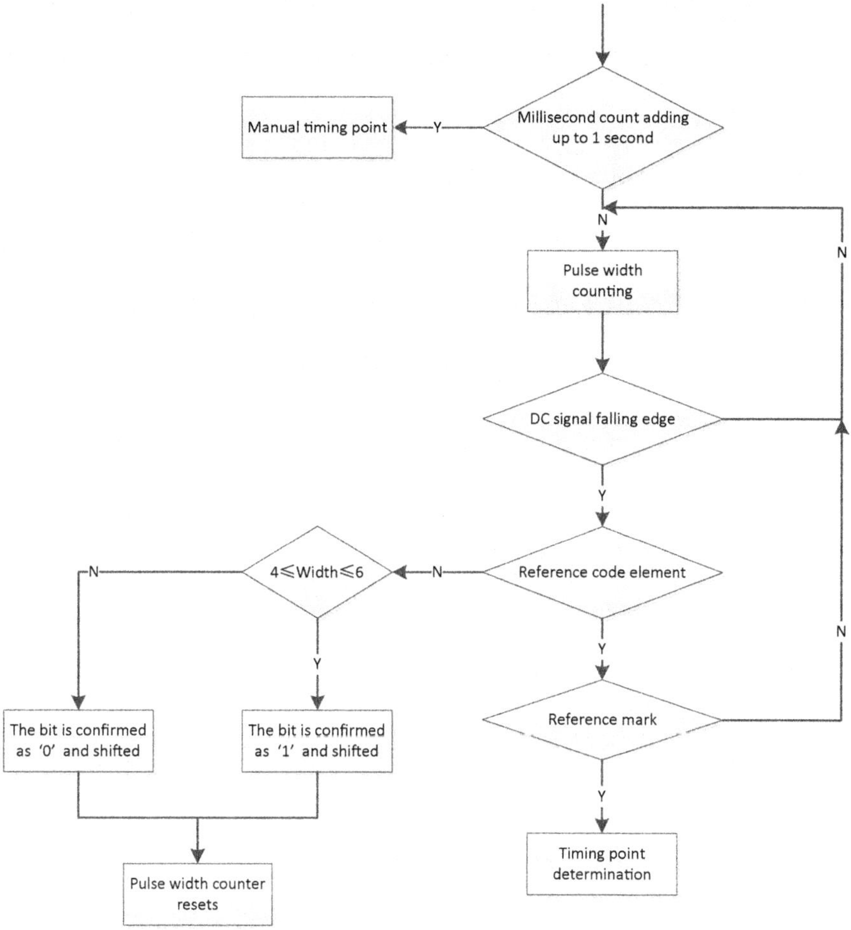

Fig. 6.17 IRIG-B code decoding flow chart

6.3.3 *Computer Functional Allocations*

Most back-end facilities are computers, including the main control computer, data processing computer(s), monitoring, data interpretation terminals, etc.

1. **Main control computer**
   The main control computer controls launch and test procedures, such as uploading flight software and element data, receiving self-test results from the OBC, starting system or prelaunch checks, and terminating some activities. It also receives instructions from higher level command and control systems to start the launch process. Current work progress, test results, and the status of the key equipment statuses are displayed on the main control computer screen.

2. **Data processing computer(s)**
   Received data are analyzed, processed, judged, distributed, and stored by computers using real-time operating systems to treat huge data volumes. They are also responsible for data interaction with other subsystems.
3. **Monitoring computer(s)**
   Key parameters representing OBS and GS operational statuses are filtered and displayed, including the following.

   (1) Subsystem test results.
   (2) Sequential instruction test results, and timing accuracy.
   (3) Connection state.
   (4) Powering state.
   (5) Operational status (similar to heartbeat signal).
   (6) Chain condition status, such as serving arm swing, payload ready, etc.
   (7) Initial states of key devices, such as IMU and ICU outputs, levelness, etc.
   (8) PS parameters, including voltage, current, frequency, etc.
   (9) Servo system parameters, including position feedback, hydraulic system pressure, motor speed, etc.
   (10) Propellant tank pressures.
   (11) Electrical leakage indicators.
   (12) Inertial meter heating temperatures.
   (13) Onboard device cumulative power-up time.

4. **Data interpretation terminals**
   Users employ this computer to browse test data, including raw and processed data, usually offline after test completion.

Computer roles have mutual penetration. For example, current work processes are also updated on the monitoring computer, so operators have details about the current situation; engineers using the data processing computer can also browse historical data for comparisons, etc. However, the four roles introduced above cover back-end computer basic requirements.

6.4 Front-End and Back-End Communication

The front-end and back-end systems are connected through the network, with switches installed at both ends, and with optical fiber as the communication media [18–20]. The network is commonly used for all facilities, so reliability is the primary design focus. Figure 6.18 shows a typical network system, comprising the network platform, command and monitoring, data service, and network monitoring equipment.

The network platform comprises four backbone switches, service tower switches, optical fiber, browser, and firewall. The platform provides network access and information transmission for each subsystem.

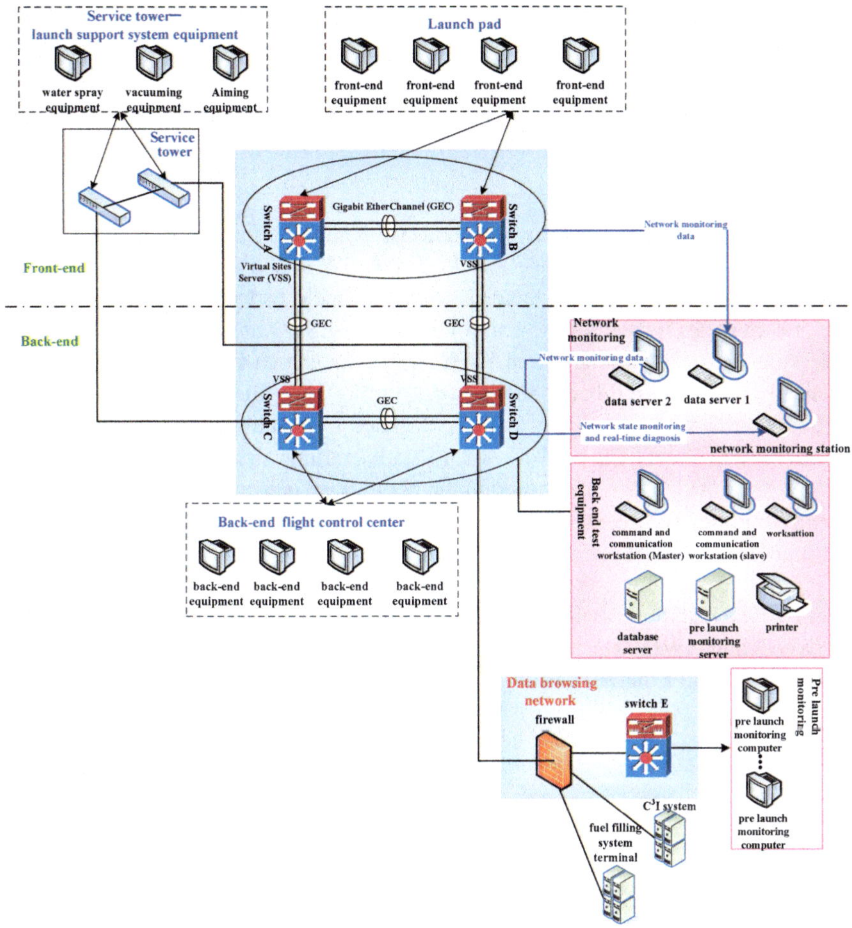

Fig. 6.18 Front- and back-end systems network configuration

Command and monitoring equipment include command and communication workstations (CCWs), a prelaunch monitoring server (PMS), and decision-making workstation. These equipments collect and display important information, provide current processing status, and actively control information exchange between subsystems and data processing.

Data service equipment include the database and prelaunch monitoring server(s), which are used for monitoring, browsing historical data, print services, and providing interface to the communication, command, control, and intelligence (C3I) system of the launch site and fuel filling systems.

Network monitoring equipment includes the data packet access server and network monitoring workstation, which monitor network equipment and command transfer status.

The following sections focus on the network platform and interface between subsystems.

6.4.1 Network Platform

Figure 6.19 shows a typical network platform, where the backbone network comprises switches SWA, SWB, SWC, SWD, SWX, SWY, and optical fiber. SWA and SWB are front-end switches located on the launch pad; SWC and SWD are back-end switches located in the control center; and SWX and SWY are located in the service tower. Browsing switch SWE allows the launch site C3I and fuel filling system to connect to the backbone network through the firewall.

The information transmission path is designed by static routing. Depending on the topology of the TLCS network, many static routings are designed in advance, arranged in priority order, and stored in the network switch configuration file. When network topology or link states change, the switch finds the correct path depending on the predesigned static routing.

To improve reliability, switches, backbone link, and network ports are redundant designed. The backbone switches adopt the VSS protocol, where (SWA and SWB) and (SWC and SWD) are mutually redundant. The fiber link between backbone switches is gigabit Ethernet and the two physical links are mapped onto one logical

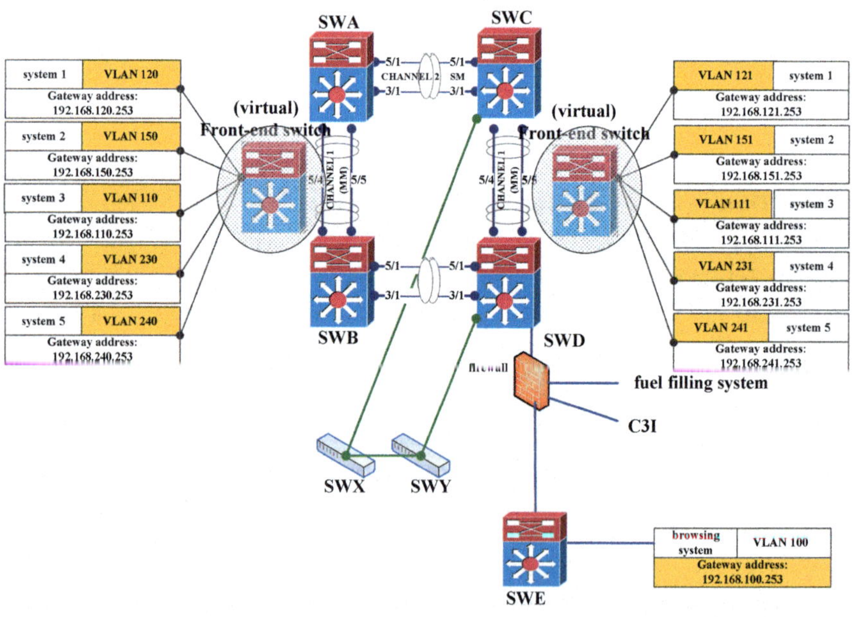

Fig. 6.19 Network platform configuration

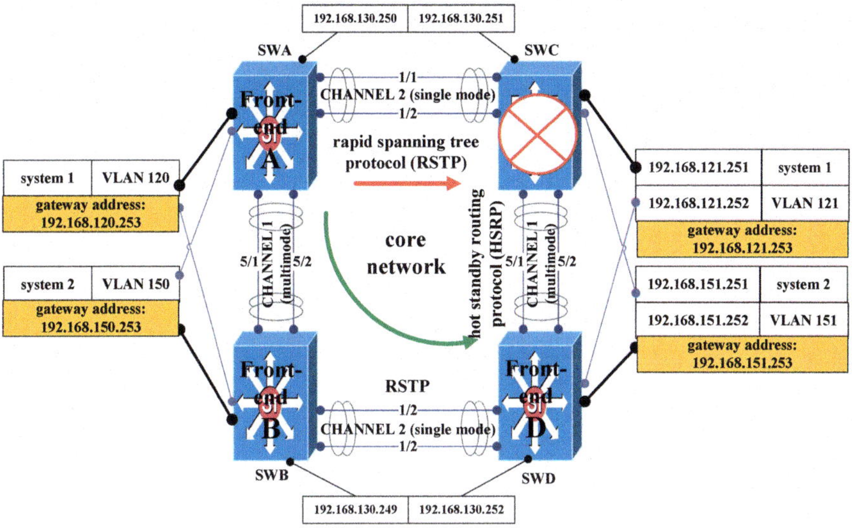

Fig. 6.20 Network reconstruction convergence process

link. This type of redundant design considers load balancing and can not only improve transmission bandwidth but also eliminates transmission single failure mode. Each network equipment's IP address is bound to two network cards, configured in TEAM mode to connect to the mutually redundant switches, so the network ports are also redundant. When a switch, fiber optic, or terminal network cable fails, network reconstruction converges quickly based on the network configuration protocol algorithm, as shown in Fig. 6.20.

Under normal circumstances, SWA and SWC complete front-end and back-end communication for system A. If SWC is out of service due to a failure, channel SWA → SWB → SWD is quickly determined by the algorithm, backup SWD is activated, and the default gateway also switches to SWD. Network convergence time is less than 1 s.

6.4.2 *Software Interface and Prelaunch Monitoring*

The software interfaces between the network system and each subsystem mainly refer to the interfaces between the CCW and each subsystem's measurement and launch control workstation (MCW). The CCW transmits commands to the MCW, receives the responses of each system, and collects relays and stores network data for each system, as shown in Fig. 6.21.

The network system is also used for prelaunch parameters real-time monitoring [21], historical data access and browsing, data storage and relay, and large screen projection. These functions run in the PMS and database server. Prelaunch

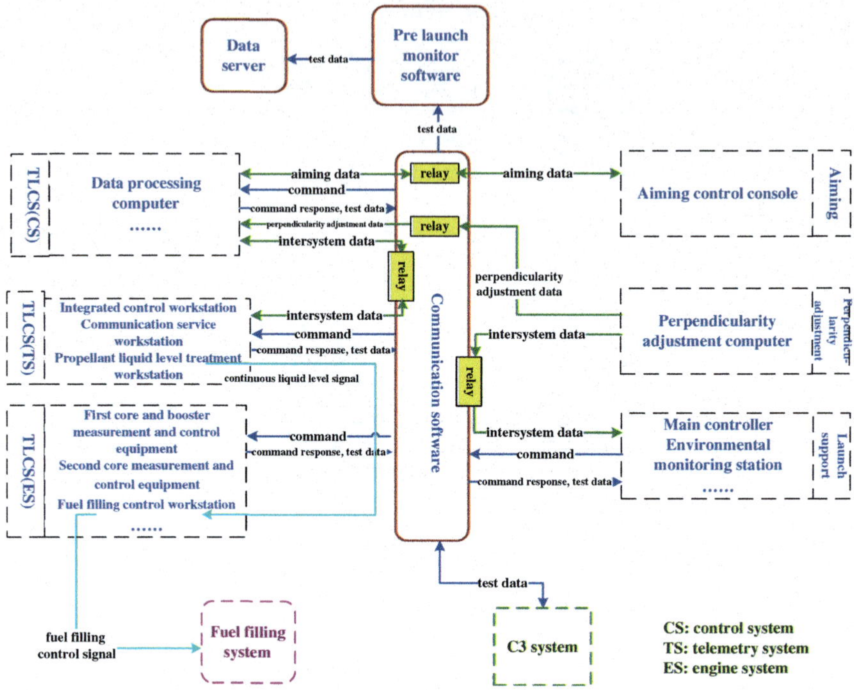

Fig. 6.21 The software interfaces between the CCWs and MCWs

monitoring software includes the data center, data source communication, and human–computer interaction (HCI) client, where communication between the client and data center is through the TCP protocol, as shown in Fig. 6.22. Prelaunch monitoring software provides multiuser browsing services based on the browser/server (B/S) architecture, and different data centers can be selected when the users open the browser.

6.5 Ground Test and Launch Control Software

Ground test and launch control software are very important TLCS components. TLCS hardware configuration is very similar for most applications, due to the adoption of platform design and productization, with the software allowing different functions.

Compared with onboard software, ground software code is very large and difficult for integration testing. Therefore, ground software development is expected to be referred as standard hardware test modules, with standardized interfaces for quick system integration and application, or even “plug and play”. For example, the US Goddard mission service evaluation center (GMSEC) proposed a standard

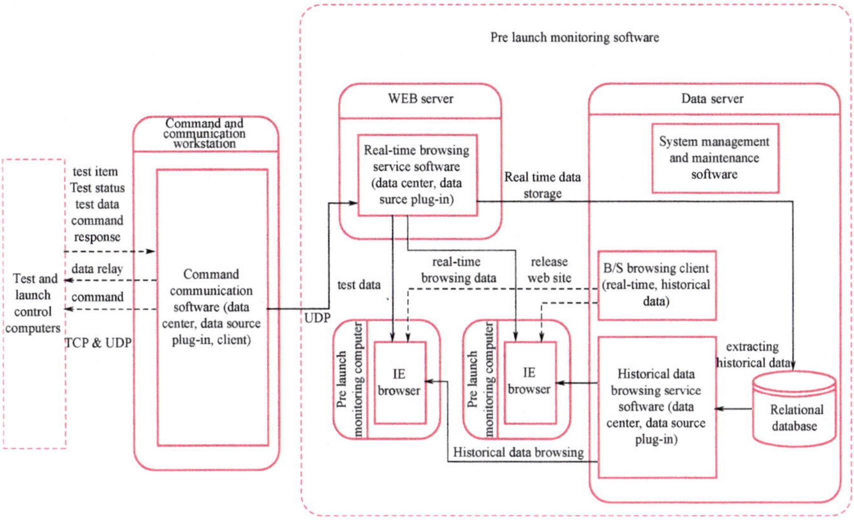

Fig. 6.22 Prelaunch monitoring softwarere architecture

architecture for ground software systems, with information flow between various systems managed by API middleware, based on a concept similar to a software bus. The aim was to allow software integration of all SLS GSs.

With increasing demand for integrated, automatic, and autonomous operation, the amount of embedded software has also increased significantly, e.g., embedded software is widely implemented in the IRIG-B terminal, various test modules, etc. Therefore, it is impractical to describe the whole ground software system, and this section focuses on the relationship between main ground software configuration items from a systematic point of view, and the key technologies unique to them.

6.5.1 Composition and Function

Table 6.2 lists the main components and functions of ground software, which vary slightly from one launch vehicle to another.

The software in Table 6.2 is basic and typically configured system level applications that interact with each other in the process of automated testing and launch control. Programs 1–8 are applied on all test sites. If the vertical test mode is adopted in the launch site, the IMU, rate gyros are installed onboard, rotating platforms are no longer required, hence, program 9 is usually configured in the integrated laboratory and assembly factory.

Figure 6.23 shows the data flow for the software system of Table 6.2.

For testing, the operators directly press console buttons or use the HCI interface to impose commands. In test mode, test instructions are sent from the master

Table 6.2 Ground software main components and functions

Index	Name	Function
1	Master computer test and control program	Core module for TLCS process control and testing. It controls PLCs and test systems, providing process control, data collection, interpretation, and transmits all raw data and analysis results to the data processing computer and display monitors. It also uses the data processing program to exchange control commands and test data with other subsystems, and receives process control command from the higher level system
2	Back-end PLC launch control program	Handles the launch control command from the master computer, or manual instructions from the control console, and forwards after logical judgment, such as power on and off, stimulation, power switch, ignition, etc.
3	Front-end PLC launch control program	Receives the logic operation results of the back-end PLC launch control program, and directly outputs the state
4	Data processing program	Stores test data and status information and displays them on multiple screens in real-time. Allows data printing and play back. Relays test data to the C3I system, and delivers process control instructions between the higher level system and master computer
5	Testing program	Receives test commands from the master computer; controls test equipment to complete the required data acquisition, stimulation, and sequential command measurement; and returns collected data to the master computer. Test equipment adopts standard bus architectures, such as VXI, PXI, etc
6	Communication between OBS and GS	Employs LVDS and RS422 interfaces to upload flight software and element data, and receives downloaded software and self-check data
7	Virtual display program	Receives data from PLCs and ground power sources, including ground power state (voltage and current), GS working state (launch or test, etc.), and displays them in graphical form
8	Power control program	Monitors PS voltage, current, and frequency; and transmits this information to the virtual display
9	Rotating platform controller program	Receives commands from the master computer, and rotates the platform. Employed to test the IMU, rate gyros, accelerator assembly, etc.

computer program (①) to the testing program (⑤), and active test work is completed by the VXI/PXI device. When the test work requires OBS cooperation, some operations are completed by the OBC, using an installed test program. The onboard test program receives commands from ① through the communication system between OBS and GS (⑥), which controls the onboard test program uploading, receiving OBC sampling results, and transmitting them to the data processing program (④). The data processing program also receives data from the testing

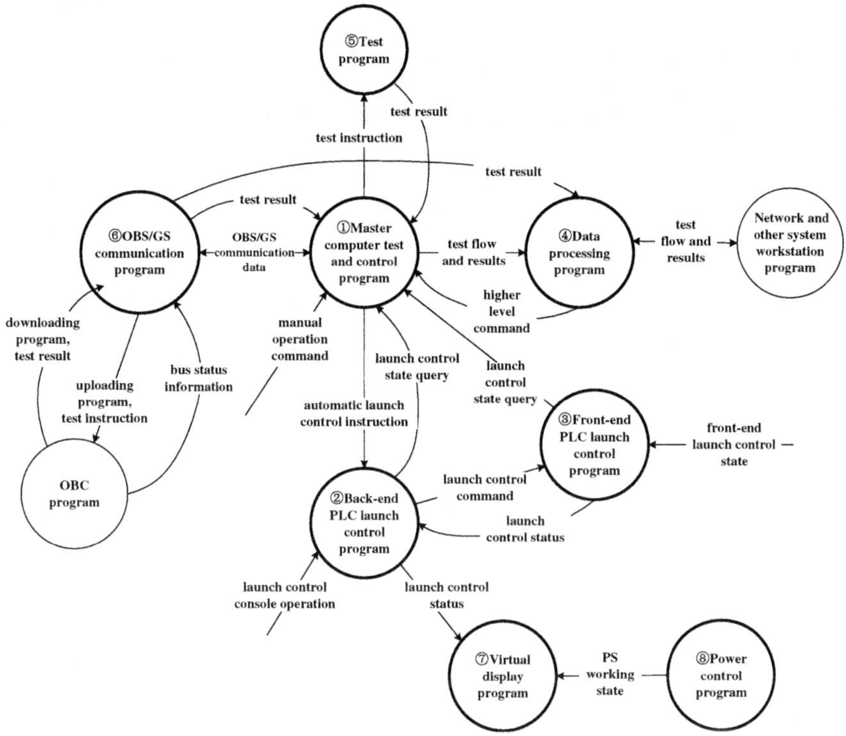

Fig. 6.23 Data flow between TLCS softwares (on launch site)

program (⑤) relayed by the master computer program (①), exchanges test data and test processes with other subsystems, and relays higher level commands to the master computer program (①). It (④) also provides data displays and distributes them to the interpretation computer for browsing and analysis.

For launching, operators press or release control console buttons, or initiate the master computer program (①) for automatic operation. These commands are received by the back-end PLC (②), and sent to the front-end PLC (③) through the network, where the commands are executed. If operators remain in the front-end site, launch control can be conducted by them directly. All launch control states are collected and sent to the virtual display program (⑦), where those states with front-end power state transferred by the power control program (⑧) are updated in real-time.

The entire TLCS software scale is very large, and the introduction above is a basic description. The communication interface and information exchanged between the software vary widely, hence when many software modules are integrated, it is difficult for them to be effectively and comprehensively verified. While a new concept allows the software system shown in Fig. 6.23 to be reconstructed into a typical software architecture incorporating standard “software bus” interfaces,

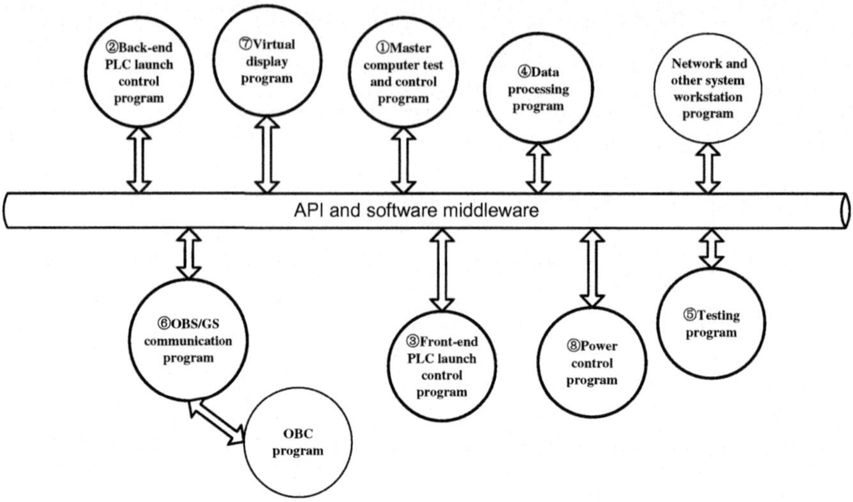

Fig. 6.24 API and middleware-based software system architecture

as shown in Fig. 6.24, and this is what the research on the ground software standard architecture aims for [22, 23].

6.5.2 *Software Reliability*

This section presents software reliability design from seven aspects, summarized from the application viewpoint. For common software reliability design best practices, please refer to [24–27].

6.5.2.1 N Version Design

N version design is an effective way to avoid common software mode failures, but increases design complexity, particularly for synchronous outputs, and procedures with complex status and numerous branches. However, for programs with relatively simple paths mainly running in sequence, or errors due to out of synchronization, which are far less than the requirements, the N version design is preferable. Launch control software related to ignition and emergency shutdown is well suited to this scheme. If N version design is particularly compulsively required due to reliability concerns, core functionality can be separated from the complicated software as a relatively small configuration item to simplify redundant design.

6.5.2.2 Signal Acquisition

1. The time delay when the stimulus is applied during the static test should be planned to ensure the signal is stable, and sampling is conducted in the stable phase.
2. After the stimulus, all equipment, including those expected to produce a response and those that are not, should be checked to avoid missed detection.
3. Filters are required for sampling analog signals.
4. State signal sampling should occur in continuous cycles (single sampling is insufficient), and be verified by majority voting.
5. Reasonable judgment is required for sampling parameters.
6. A time threshold to allow sampling should be set when sampling sequential signals, to avoid misjudgment of the next signal under coupling interference conditions.

6.5.2.3 Data Communication and Exchange

1. Regardless of the protocol adopted in the underlying hardware, reliable communication protocols should be designed at the application layer.
2. Response modes are required for communication of important signals.
3. A cyclic queue shall be provided for the no response mode of communication or data exchange, to avoid previous data being overwritten by subsequent data.
4. Asynchronous communication read and write buffers should be considered to avoid conflicts while simultaneously reading and writing.
5. A timed exit mechanism is required when using the query method for communication, to avoid endless loops when the checked state fails.
6. The semaphore-based data exchange system shall avoid interlocking.
7. Shorter segments are preferable for reliable transmission, where the retry mechanism and error warning after certain retries should be set. Both methods aim at effective processing under the failure conditions.
8. Communication should not be interrupted by higher interrupts, and the code should try to ensure all data are written to the buffer once, e.g., data written in a single cycle instruction will not be interrupted.
9. Each data frame should be accompanied by a time tag to ensure that the recorded data is meaningful.

6.5.2.4 State Output Control

1. Actual output results for state control signals with strict timing requirements should be checked, and the next step should only be performed once the current instruction has been correctly conducted and verified.

2. Level signals are preferred than pulse or edge triggered signals. If power consumption is not a concern, the effective valid signal time should be as long as possible;
3. The current response should be used as the starting point for the next control step. This will avoid differences in the hardware response time, which may cause unstable states or sneak circuits; and this can also make the operation irrelevant to the hardware.

6.5.2.5 Troubleshooting

1. There should be clear instructions to deal with fault cases, and manual intervention must be configured to stop the process, ignore the failure, or continue.
2. Detailed fault information and real-time alarms can be produced by different task levels; only the alarm is required in processes with high real-time requirements. The details can be stored in the background or browser terminals to avoid affecting the main process.
3. Key parameters and warning parameters should be treated differently.

6.5.2.6 Real-Time Operation Requirements

1. Select a controllable timing system that is independent of the operating system and CPU.
   In the ground computer, some timing instructions are related to the CPU frequency, but the computer can reduce the frequency in the case of excessive heat dissipation, resulting in inaccurate timing. Therefore, timing should be based on instructions unrelated to the frequency, or on trusted external source(s).
2. Instruction with definite temporal relations must be handled in a RTOS.
   Non-RTOSs cannot be applied for these instructions due to their inherent timing uncertainty. To solve this, a real-time kernel can be integrated into the non-RTOS; or the upper and lower computer architecture can be adopted, where the RTOS runs in the lower computer.
3. Timing cannot be based on execution time of high-level language instructions.
   High-level language is often compiled into a number of assembly instructions, with varying execution time under different conditions. Instruction prefetch is also commonly used in many compilers or CPUs, which makes execution time even more uncertain. Optimization options during compiling will also lead to different target codes, and affect execution times accordingly.

6.5.2.7 Redundant System Testing

1. To ensure test coverage and real-time requirements, the GS should set relatively independent parallel processing programs to interact with the onboard redundant system, such as multi-threads. Serial processing of redundant systems by a single program or thread is not recommended here.
2. Software runs asynchronously on the OBS and GS, and this will affect information processing of redundant components. Therefore, a buffer with sufficient capacity shall be set aside to ensure the data would not be lost even if the response is not made in real-time;
3. If an onboard TMR function is evoked by the ground signal, when the two onboard modules are synchronized and enter the next state, synchronization mechanism is generally released, causing the third module to remain in the previous state, unable to reach the same status as the other two modules. Therefore, the GS should control the synchronization mechanism to release after a delayed time, or after receiving all state transition responses from each redundant module.

6.5.3 Process Control Language and Script Files

Demand for a special launch control and test language has been very strong, to adjust and reconstruct the operational process flexibly on-site; and improve reliability, reusability, and adaptability of the ground software system. NASA first proposed the "acceptance test or launch language", and subsequently developed the "ground operations aerospace language" [28] that has similar grammar to natural human languages and related operations, and is used to arrange test and launch control commands. Many different testing and launch control languages have emerged worldwide, including "system test and operation language" [29], "space station operation language" [30], and the "spacecraft command language" [31] developed by Interface Control Systems, Inc. Europe has also produced "space control and operations system II: operation language" [32], where system engineers can edit, test, validate, and install scripts written in the operation language with no requirement for software support engineer involvement.

However, these languages are highly customized and lack versatility or portability. COTS software-based script files for scheduling, which must be compiled to run, have become more common to control the operation. For example, the "enhanced and redesigned scripting" [33], is very effective for flight control operation, and can be easily integrated into the GS; A variety of functions are provided, including process control, with both C# and Visual Basic programming languages available.

Each Chinese LV has its own TLCS and software system, with operations and processes fixed on the specific launch site. Hence, there has been very little call for changes, and the demand for a scripting language is relatively small. However, Chinese manned LVs gradually realized the conveniences of separating test process and software, due to requirements to change testing processes for troubleshooting and subsequent improvements. Thus, developers began to use PCL and script files to design the GS software [34]. PCL schedules software operational conditions following grammatical rules that facilitate instruction parsing, and a script file describes the scheduling instructions. With increasing demand for reuse, this approach has gradually become the main technical solution for GS software design.

To facilitate this trend, some TLCS software is encapsulated into the minimum program unit with standard interface. The main control software analyzes the description in the script file and calls the relevant procedures in order to complete the operations. The script files can consider all possible combinations of operational items through the process, and operators can be authorized to add new items as required. Script files are in text format, to facilitate editing on-site. Some provide graphical interfaces, and web-oriented scripting languages have emerged to boost cross-platform applications [33].

These solutions increase software adaptability. For example, the TLCS can control power-on for the OBS, by assuming item 1 is powering on inertial devices, item 2 is the OBC, item 3 is other ICUs, and item 4 is the SCU, etc. Each power control item is conducted by the same program; only the control parameters differ to meet the different requirements. These items are encoded and stored in order in the script file following PCL syntax. When the file is called the main control software analyzes the meaning of each item description and performs the corresponding power-on operation. Under some conditions, not all devices are powered, such as when testing the sequential control subsystem only items 2 and 3 are required, so the operator can modify the script file, adding a statement related only to these two items, and call that statement to commence the test.

We briefly introduce the technology applied in Chinese TLCS with technical details provided in [34]. Following the above concept, TLCS minimum program unit functions include at least power-on and power-off, state setting, status query, test control, data processing, and communication. The operational item must cover subsystem tests, integration tests, general check, and launch control, etc.

Figure 6.25 shows the typical script file format, where various operations can be described depending on the grammar rules. Each operation is considered a record, so a script file is a collection of N records that may change based on the test requirements. Each record describes a test or launch control operation independently, including several fields to clarify specific operational requirements. The number of fields and their specific meaning for each record can be defined as required, and the definition represents the syntax of the script control language.

Field length and definition can vary depending on the type of operation. For example, the first field of each record can be defined as an operation identification code (OIC), as follows.

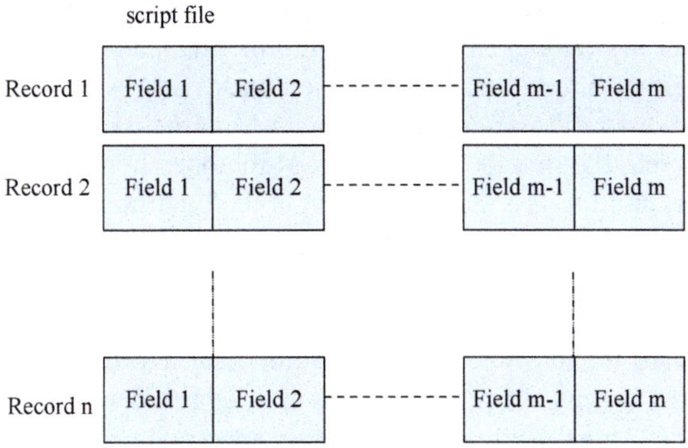

Fig. 6.25 Typical script file format

1. OIC 0: represents the test data processing class, and subsequent fields represent calculation functions, parameters, etc.
2. OIC 1: represents the test program control class, and subsequent fields represent corresponding elements in the data structure of the measurement module.
3. OIC 2: represents the communication class, and subsequent fields mainly meet RS422, RS485, PLC SRTR, TCP/UDP, etc. communication requirements.

The main control software first loads the script file, transfers all the records into the computer heap space, using an array of pointers relating to each record, forming an index, as shown in Fig. 6.26.

Once the script file is loaded successfully, an array or linked list is generated to store the operational item numbers (OINs). Each operational item, such as items in the subsystem test, or power-on items in the general check, etc., is composed of

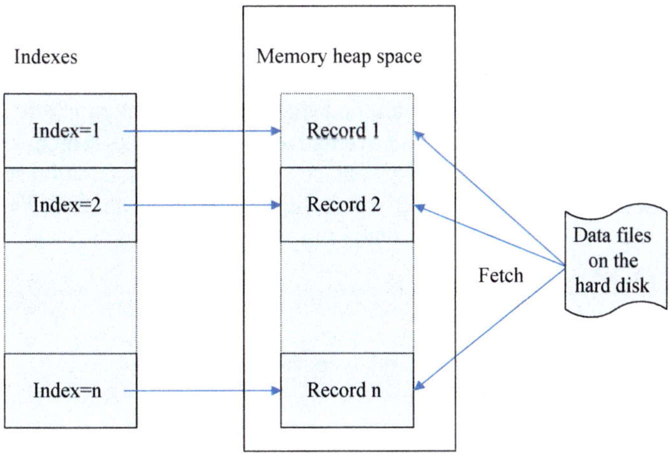

Fig. 6.26 Script file loading

several records and is used for operational control and test data processing. The operator can enter the OIN in any combination, the main control program searches the records corresponding to the OIN, and then parses and executes the operational item's records in order. The search is sequential from the beginning of the file. All records are browsed, the first record location of the operational item is recorded as the index header, and the last record location is the total number of operations for this operational item. These data are relayed to a file parsing function.

Script file for test and launch control processes significantly decreases TLCS software code volumes, while expanding their data volumes, which is critical for improving software reliability. Software development and maintenance efficiency are also improved if only script file records are regulated without modifying the program unit code. This flexible implementation approach means new script files or new records can be easily added to meet expanding TLCS requirements.

6.5.4 Prelaunch Elements Generation and Quality Assurance

There are more than 2000 launch elements within the FSW. For satellite launches, elements are verified approximately one month before launch and are checked on-site. However, for rendezvous and docking missions, elements related to the target orbit (spacecraft) and aiming are determined only four hours before launch, and traditional manual element preparation cannot support the real-time requirements. Therefore, a system that can quickly generate the elements and assist with quality assurance is essential.

Launch element generation compiles various original parameters into data the FSW can identify and use. The compilation itself does not require much time, but the original parameters and generated object code must be checked to ensure compilation correctness, and the object code must integrate with the FSW to be verified in the form of HWIL or SWIL testing.

Using the system introduced below, FSW (including launch elements) can be evaluated on the launch site, under all kinds of disturbance combinations and fault simulations, and launch element test coverage approaches 100%, which is regarded as the basis for FSW delivery. The advantages of automatic test configuration, and template-based data acquisition and report generation, mean that the system can also assist with unattended 24 h continuous simulation to apply sufficient stress testing for the FSW.

6.5.4.1 Launch Element Types and Generation Processes

The launch element generation is the transformation of the raw data into a resolvable format, which are required flight control parameters, as shown in

Table 6.3 FSW element parameters

Serial number	Category	Details
1	Guidance control parameters	Physical constants, state description parameters, cutoff thresholds, cutoff time, perturbance guidance parameters, iterative guidance parameters, initial aiming parameters, integrated navigation parameters, IMU redundant management parameters
2	Attitude control parameters	Rate gyro compensation coefficients, control law gains, program angle filter parameters, guiding parameters, correction network parameters
3	IMU parameters	Meter equivalence, error calibration coefficients
4	System parameters	Flight time and corresponding velocity, position, pitch and yaw program angles; aiming parameters, target orbit parameters
5	Sequence parameters	Cutoff permission setting, timing cutoff setting, etc.

Table 6.3. The elements are divided into two categories: data determined one month before launch, including guidance parameters, attitude control parameters, IMU parameters, precise trajectory; and those verified just several hours before launch, such as target orbit parameters, aiming parameters, etc.

The aiming scheme is related to the inertial devices. If an inertial stabilization platform is employed, aiming is achieved by imposing moment on the gyros, rotating the platform body, and strictly aligning the aiming prism to the fly plane. Since the theoretical fly plane direction is determined in advance, launch azimuth is not considered an element parameter.

When the IMU is installed onboard, only the LV body can rotate to make the IMU aiming prism strictly align to the fly plane. After fuel filling, it's not safe to turn the LV body through a large angle, or rotate it frequently, so the actual aiming azimuth is measured and uploaded to the flight software. This is then used to control the LV roll after liftoff, redirect it to the theoretical azimuth. These aiming elements, such as the angle of elevation and alignment azimuth, should be surveyed just before launch.

The spaceship target orbit changes with the targeted spacecraft's orbit fluctuation in rendezvous and docking missions. To reduce fuel consumption for orbital adjustment, the later the target orbit is determined, the less the error caused by target orbit prediction. Target orbit parameters include perigee, and apogee altitude, argument of perigee, orbit inclination, and RAAN, etc.

The element generation flow is shown in Fig. 6.27. If the target orbit parameters are determined just hours before liftoff, launch elements must be generated twice, i.e., the guidance and attitude control parameters, and standard trajectory parameters are derived earlier and tested; whereas the aiming and target orbit parameters are integrated at the last possible moment and verified using well-directed test cases. Since the FSW is usually programmed in a high-level language, an inverse transformation should be performed to check the generation; then, some SWIL or

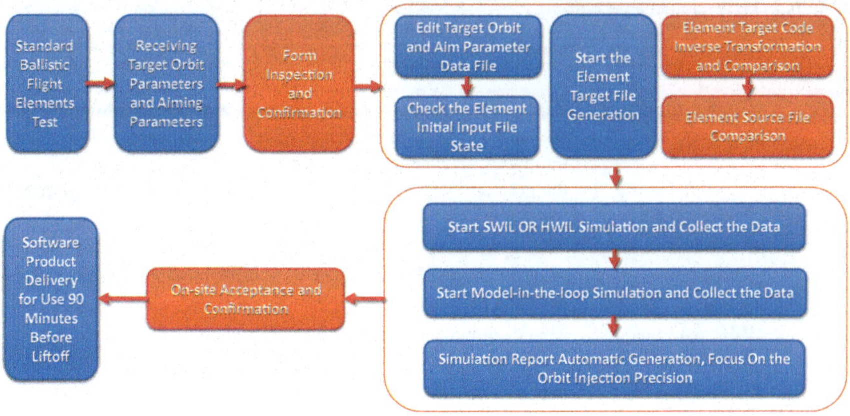

Fig. 6.27 Element generation process

HWIL simulations performed. When all checks pass, the final version FSW is verified for delivery.

6.5.4.2 Hardware and Software Design of the Generation System

The element generation system hardware design focuses on HWIL/SWIL simulation, as shown in Fig. 6.28, referring to the FSW simulation test platform. The flight simulation computer simulates LV motion, and generates the required physical signal for the OBC. The simulation software also samples the attitude control signal output by the FSW. Morever, a set of interface equipment or signal conditioning modules is required. The FSW runs in the OBC and its operation result is recorded by the telemetry interface for analysis, and a telemetry processing computer is set up for this purpose, which also plays a role in program and element uploading. An upper computer for simulation and telemetry computers, called center computer, manages and schedules the whole process, including launch element editing, compiling, comparing, and HCI. If it is a SWIL simulation, only a simulation computer is required.

Depending on the hardware design, software functions can be divided into a central computer, simulator, and telemetry software, with the software framework as shown in Fig. 6.29.

The important role of the central computer software is to generate the launch elements (converted from the raw data), compile the source file of elements (from source into object code), make some comparisons, and complete element review reports. These functions are integrated and regulated by a unified resource management work console. Due to the high security level for element generation and compilation functions, they can be designed as independent software configuration

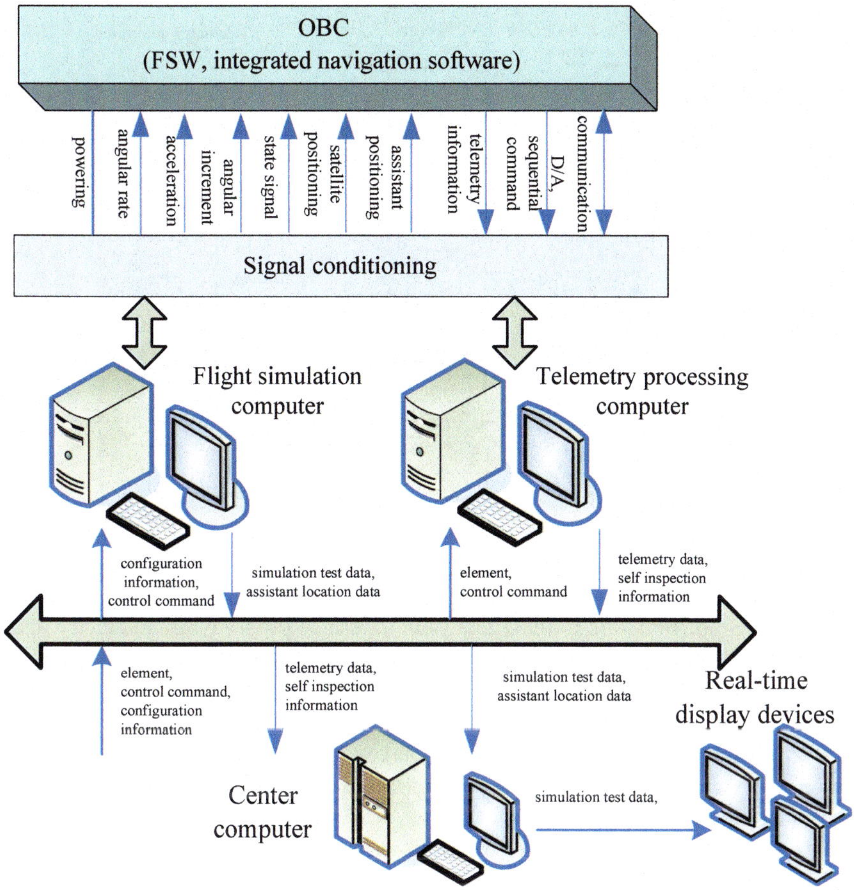

Fig. 6.28 Element generation system

items and called by the console, which is conducive to quality control. This console also manages documents generated in the process, such as data files, configuration files, comparison reports, etc.

The remote-control console on the central computer controls and configures the simulation and telemetry computers to complete data file transfers and instruction communications, through a communication agent, such as transmitting control instructions, receiving responses, downloading configuration and target code files, receiving and storing collected data, etc. The agent assures each computer software's independence, to a great extent.

The third function of the central computer software is to display the collected data, which is integrated into the data sampling and display console.

For HWIL simulation programs, the mathematical simulation program and GPS auxiliary data receiving program run on the simulation computer, where the latter

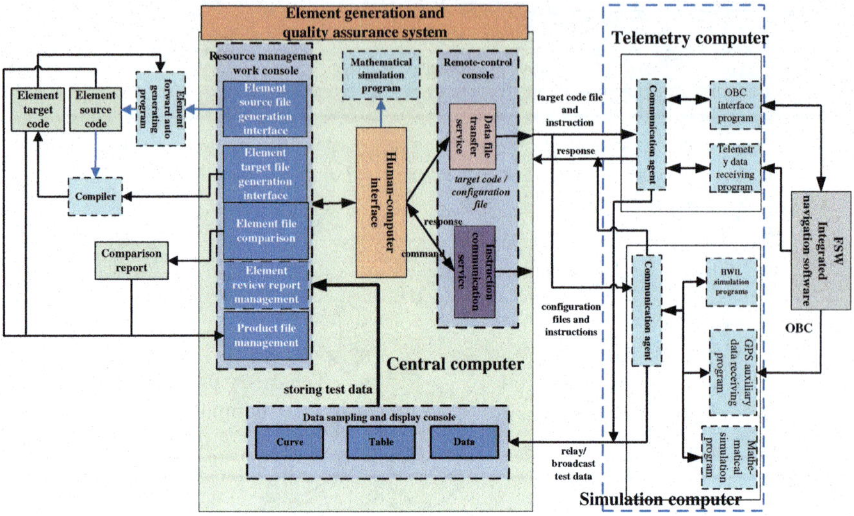

Fig. 6.29 Software framework of element generation system

receives GPS position data frames. The OBC interface program and telemetry data receiving program run on the telemetry computer, and the former is used to upload flight control software.

Launch element comparison is an important quality control checkpoint. The comparison transforms the target code to source code, and compares the generated source code with the initial file to confirm the generation and compilation. The process is shown in Fig. 6.30.

The inverse transformation uses map and variable template files. It analyzes the map file to obtain base addresses of all variables and arrays in the data section of the object code, and the size of each variable or array by analyzing the variable types in the template file, and then captures the specific object codes. Once captured, the content is converted to decimal code usually used in high-level programming language source code, until target code analysis is completed. It then analyzes the source code content, and based on the variable name and "=" operator, matches the decimal code with the variable name, forming a new source code file.

The new and original source code files are compared line by line, depending on the variable names in the template file. If the variable is not a floating-point type, then variable assignments are compared directly; if it is a floating-point type, a segmentation analysis is done depending on the decimal position, and errors before and after the decimal position are calculated with reference to user specified error thresholds. If the error is within the threshold, the two variables are regarded as equal, otherwise they are considered unequal and marked as a deviation. The comparison continues until the file ends, and the code notes are filtered. If there is no deviation, the element generation is verified as correct; otherwise the deviation causes should be checked.

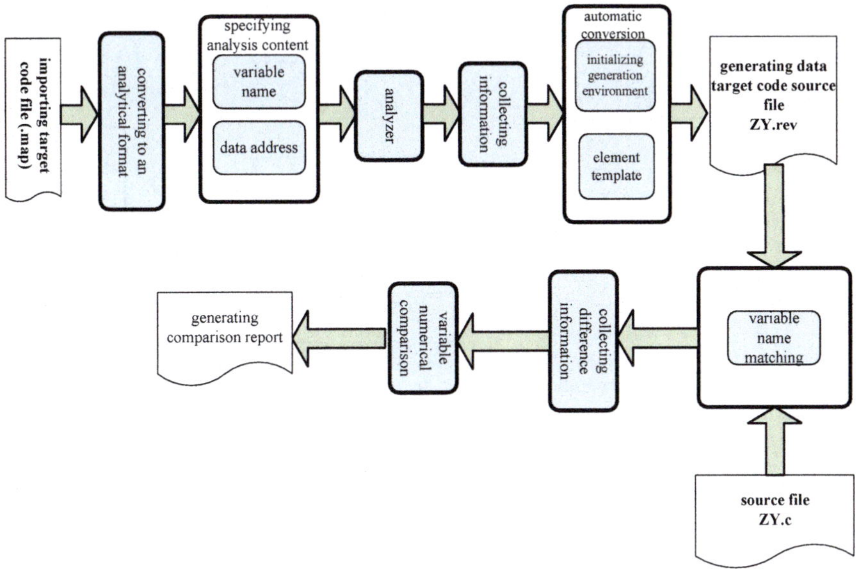

Fig. 6.30 Element comparison process

Thus, launch element generation and comparison are achieved automatically and quickly, with assured accuracy. The acceptance report can be generated automatically on-site using data file template. This not only ensures quality, but also improves operational efficiency.

6.5.5 User Interface Automatic Generation

The graphical user interface (GUI) provides user feedback during operation. With the development of information technology, users not only make higher demand for software function, but also higher GUI requirements, such as esthetics, customizability, etc.

GUIs can be developed using auxiliary design tools or direct coding. Using design tools, the GUI developer does not need to master too much technical details, and the layout can be more intuitive, but can also be difficult to expand and customize, and the code is hard to reuse, debug, and maintain. Using direct coding, the developer can easily extend the controls, or customize a new control, and useful controls can be reused, but the developer must master significant technical details, and layout and coding is hard work, with relatively long development cycles. In practice, the GUI designer needs an art foundation, to skillfully apply graphical editing software, which increases the development difficulty.

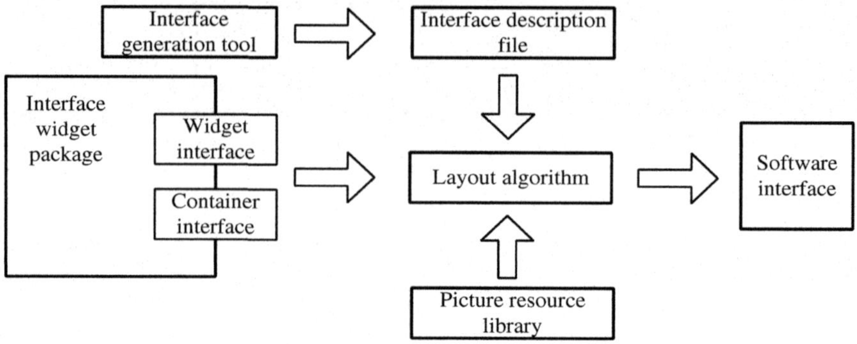

Fig. 6.31 Automatic GUI generation process

Therefore, automatic GUI generation has been proposed, as shown in Fig. 6.31. The concept abstracts the GUI into three related elements.

1. Control
   Control is the abstract of the visual elements on the interface that have independent function and/or logic. Each control has a set of attributes and executable actions, e.g., a separate button, or two related buttons can be abstracted as a control (widget).
2. Container
   The container is a window or panel containing a group of controls with related functions or logic, basically a special control type. For example, controls used to display a device's parameters can be placed in a container, which can be hidden until required. Each container can include any number of controls and sub-containers. Containers can also be abstract and have no corresponding visual interface elements, which is useful for grouping related controls.
3. Layout
   The layout element implements the mapping between a control and a relative location on a container, and establishes a dependency relationship between the control and the container, and the overlay relation of different controls at the same location. The interface of any application can comprise a master container with sub-containers and controls contained within it.

The control and container interfaces package the graphical interface elements or their combinations. The control interface provides methods for accessing control properties, defining control actions, and creating control instances. The container interface defines methods to manage child controls, such as inserting a control, deleting a control, or searching for a control. Since the container is a special control, each container also implements the control interface as well as the container interface. In general, the method for creating a container also creates instances of child containers and child controls. The layout algorithm parses an XML format interface description file at the start of the program and dynamically creates the program's GUI.

6.5.5.1 Interface Description File

The interface description file describes and defines the GUI, in the following sequence.

- Define the main container, including the container type and attributes, child containers and controls, and layout information for these within the main container.
- Define child controls contained within the main container, including the type and attributes.
- Define child containers contained within the main container, including the type and attributes, any further child containers and controls, along with layout information in the upper level container.

6.5.5.2 Interface Layout Algorithm

The interface layout algorithm inputs the interface elements and interface description files, and forms the specific GUI. The steps are shown in Fig. 6.32 and summarized as follows.

1. Extract the description of the main container type and create its instances depending on the control type from the interface description file.
2. Set the main container position coordinates and properties from the layout and property information in the interface description file.
3. Create the child control and container instance positions relative to the main container, and set their properties using the information in the interface description file. If a child container is created, its child controls should also be searched.

The GUI is complete when all the child controls and containers have been traversed.

The interface automatic generation tool provides visual interface editing, where the programmer can directly configure the GUI using mouse drag-and-drop in a WYSIWYG fashion, which is captured in the interface description file. A picture resource library contains commonly used images for aerospace test and launch control software interfaces, classified under different interface styles. The interface style can be set in advance, and the tool automatically extracts the corresponding images from the library and presents in the preview, then copies them to the product folder. These images and the generated interface description file provide the input to the interface layout algorithm.

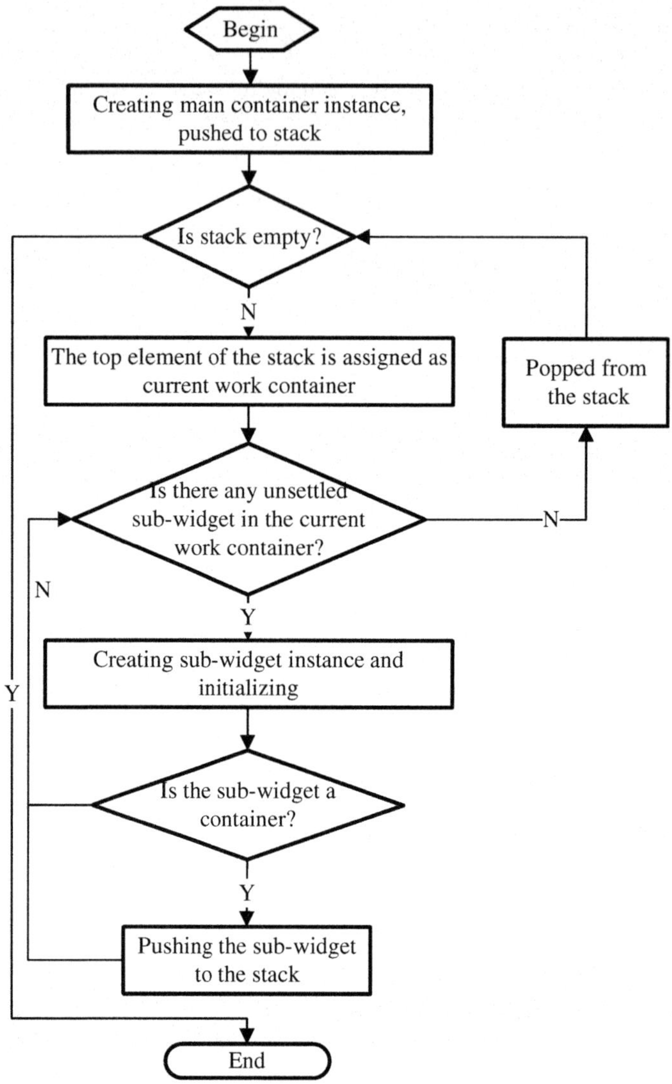

Fig. 6.32 The process of GUI layout algorithm

Fig. 6.33 An example of automatically generated GUI

Figure 6.33 shows an example of the final GUI.

6.5.6 *Software Testing*

Ground software can be very large, and software testing involves multiple systems. For example, to evaluate ground command and control software for NASA's new generation spacecraft (MPCV) and LV (SLS) mission [35], a complete joint verification environment comprising vehicle management systems, SLS mission, and fault management subsystems, etc., and participants include crew, mission control operators, launch control operators, etc., was constructed. Unlike traditional system integration testing, professional operators from manufacturing and testing departments participate to ensure independent evaluation.

It is expensive to build such complex systems to test ground software. Therefore, NASA developed the GS development and operation tools, which establishes and verifies system level model for each software module. This is an effective means of testing software interfaces and protocols between complex systems. For further susbsystem software testing, some or all software models, depending on the scales, can be replaced by real programs or code, so combined model and code or code only testing can be performed. In other words, ground software testing is hierarchical, with each level of testing having a different focus. Higher level tests focuses more on system interfaces and protocols, whereas lower level tests focus on checking individual software codes.

The test and launch control software is only one part of the ground software, and is evaluated using a combined model and code scheme. Before introducing test solutions, we review software testing techniques first.

6.5.6.1 Commonly Used Software Testing Techniques

This section expands on software testing classification introduced in Sect. 2.2.4 [36–39].

White-box testing is structural or code-based testing, where the tester analyzes the source code to design test cases, and can examine the software in detail without plug-in code. The tested software is transparent to the tester and it is relatively simple to obtain test coverage.

White-box testing is particularly significant for high-level programming language applications, where source code statements and branches may differ from the target code, so source code-based structural coverage analysis cannot effectively reflect target code coverage. Although statement, branch, and modified condition decision coverage (MC/DC) may approach 100% in source code testing, some instructions and branch jumps in the target code may still not be executed, such as structural changes caused by compiling strategies, unused library functions, etc. Only white-box testing can effectively test these target codes.

It is relatively easy to design test cases using white-box testing, because tested code is executed dynamically in the white-box testing environment, and the execution process can be observed. These data can be fed back to design the test cases, and source code and target code test coverage can be easily collected to supplement test cases for unchecked code. However, although 100% path coverage is the ultimate aim, this would require an enormous number of test cases, and is seldom used for practical evaluations. Branch and statement coverage is widely used, but they still cannot reflect the overall software logic. Therefore, other coverages are employed as supplements, such as Boolean operator coverage, which considers "OR", "AND", "NOT", and other operations; relational operator coverage, which considers <, >, =, >=, and <= operations; loop coverage, which considers "FOR", "WHILE", and "DO-WHILE" cycles, and MC/DC coverage.

Black-box testing is a type of functional or specification-based test, focusing on software function, i.e., the reaction to a given condition and input signal. These functions can be extracted from the software specification without requiring a detailed study of the code. This test is usually performed on the target machine and cannot track the execution of each statement.

During black-box testing, the tested software can only be judged by its output, and it is difficult to derive coverage information. Since it is difficult to understand the detailed software design, or the code internal structure, black-box testing case design has attracted a great deal of research interest. Test case design for black-box testing should focus on the specifications and task descriptions (extracted from a suitable file). If these are clear, such as written in a formal language, then test cases can be automatically generated [40].

Some intelligent algorithms, such as the genetic algorithm, can also be employed for automatic test case generation [41]. A set of input data is selected to evaluate the software response, and the goal of inheritance or crossover operations is to cause performance deterioration, or design margin reduction, ultimately to explore the software performance limits. This is also regarded as the objective of black-box testing. Test validity should be evaluated once testing is complete, usually in a model-based manner [42], which differs greatly from test coverage statistics in white-box testing.

Gray-box testing addresses deficiency of first two approaches, i.e., it simplifies test case design, and focuses on properly understanding the detailed software design to ensure the test cases are more targeted. Tested software runs on the target system, and the processor address and data lines signals are monitored. This blurs the boundaries between white and black-box test techniques. Combined with the code analysis and plug-in code executing state, test coverage can also be assessed.

For TLCS software, only those modules related to launch control are safety concerned, and few are embedded software. Most software runs on the Windows platform, and black-box testing is appropriate and commonly employed. Therefore, this section introduces black-box testing techniques for ground software.

6.5.6.2 Overall Scheme

Ground test and launch control software usually runs in a network environment, so some software configuration items (SCIs) can be combined for testing, depending on their level of coupling, which can simplify the test platform design. Figure 6.34 shows an example that integrates the main control, data processing, and virtual display computers. SCIs outside the box can be modeled to meet the test platform requirements. The test environment is generalized and integrated, and can simulate many different equipment. Basic modeling requirements include the following.

1. Runs on Windows or Vxworks platform as a server application, and the listening port can be modified.
2. Each test server can deploy one or more sets of simulated equipment (models) that bind different ports.
3. Test environment/software simulates real bidirectional (command and response) or unidirectional (transmitting or receiving) network communication packages, and can configure the way the data is sent, e.g., normal transmit and transmit with configurable delay, etc.
4. Data sent by software under test to simulated equipment (models) is recorded in the original HEX format, but saved and displayed in parsed readable text format.
5. Simulated equipment (models) provide user interfaces for various network communication requirements, such as single or continuous transmit, abnormal

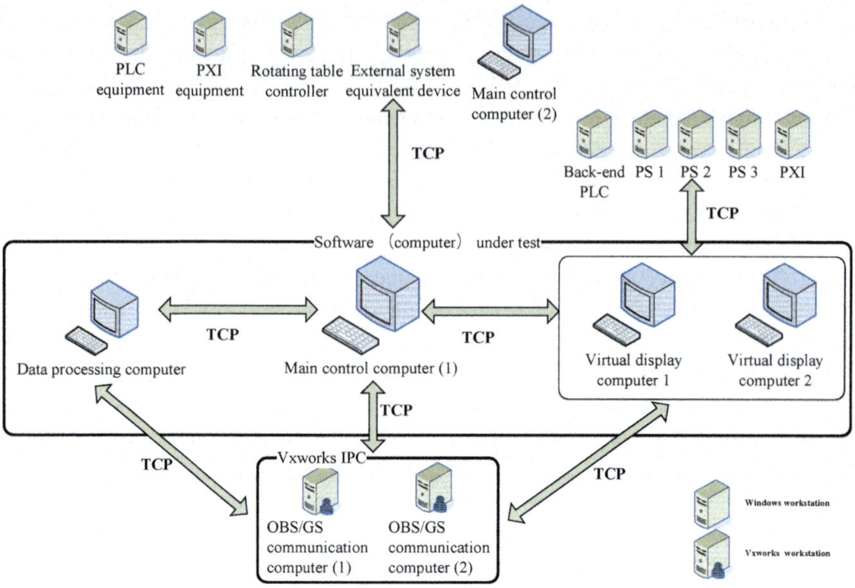

Fig. 6.34 An example of ground software test system configuration

packet simulation, etc. Abnormal types include: frame head, frame tail, data area length, checksum, and custom defined exceptions.

6.5.6.3 Integrated Test Environment

The models, or the simulated equipment, are integrated into the test platform, and public functions are extracted, such as data modeling, data modification, communication simulation, and information display. A software test environment comprises models, and service, display, callback data processing, background data processing, logfile saving, communication, and management functions.

Service

Entrance to each device is via the service engine, mainly to provide socket monitoring, analyze data received, respond to commands depending on the configuration file, and notify the GUI module for related operations. The main procedure features are as follows.

1. Multiple thread management.
   The client and server link is processed by an independent thread. Before the client disconnects, the thread records the communication status and provides corresponding feedback for the client. A new thread is started whenever a new

client sets up a link with the current device, and new service is provided for the new connection. GUI processing is also served through individual threads to ensure independence of response transactions and interface display.

2. Multiple IP binding.
   The service engine provides multiple IP binding functionality, using different threads to ensure stable and independent operation.
3. Client management.
   Clients connected to current devices are managed uniformly, making it convenient to check IP addresses, port details, and provide a manual disconnection function.

Display

The GUI is independently designed depending on the display function requirements. Communication between GUI and response processing module is through a public communication module to ensure independence. Thus, if the GUI program is in an abnormal state, this will not affect response functions.

Call Back Data Processing

A new thread is started when the socket receives a data packet, and sends the data directly to the callback data processing module, which implements the received packet storage, CRC verification, packet parsing, callback data construction, exception simulation, frame closure, callback data transmission, etc.

Background Data Processing

Background data processing provides special processing for some equipment, such as rotating platforms, and OBS-GS communication simulation. This not only avoids interference with normal command-response processes, but also ensures the accuracy of these special operations, particularly for near real-time processing, because the independent thread can ensure more accurate timing.

Log File Saving

The system provides opportunity to save detailed logs during processing. From original link establishment to disconnection, the logs include: client link information, original data packet, parameter analysis results, parsed data packet content, callback data construction information, callback data content, disconnection information, abnormal information, etc. To improve callback data feedback efficiency, the log file is saved after the callback data is transmitted. Log file size is checked after each save event, and if it exceeds a threshold, the log module automatically creates a new log file to ensure efficiency.

Communication

The network communication module simulates bidirectional (command and response) or unidirectional (transmitting or receiving) communication, as shown in Fig. 6.35 for the bidirectional case.

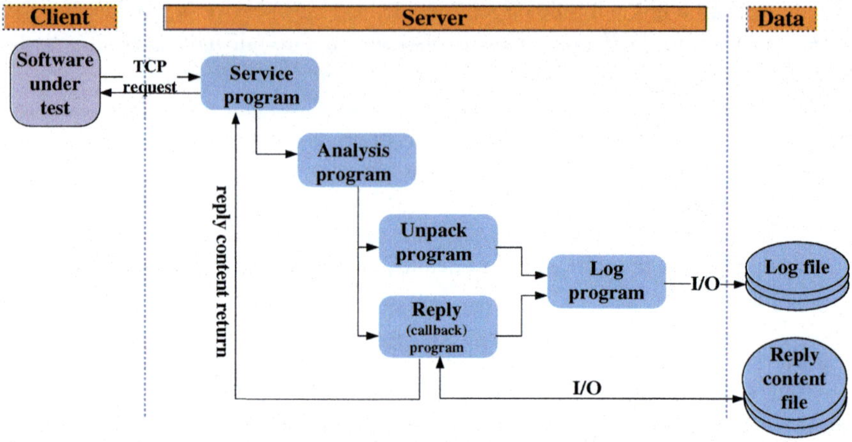

Fig. 6.35 Bi-direction package transmit process

References

1. Xu YW (editor in chief) (ed) (1992) Control system (3). Missile and aerospace collection—liquid ballistic missile and launch vehicle series. China Aerospace Publishing House, Beijing (in Chinese)
2. Song ZY, Zhang L (2007) Development and prospect of Chinese launch vehicle testing technology. In: 1st high-level forum on national defense science and technology industry experiment and testing technology, Beijing, China, Sep. 21–22 (in Chinese)
3. Zhang L (2004) Ground test launch control system of LM-2F launch vehicle. Missiles Space Vehicles (1):34–37 (in Chinese)
4. Kania K, Hill J (2011) The application of software safety to the constellation program launch control system. In: International conference on software quality, San Diego, USA, Februrary 8–10
5. Lougheed K, Mako C (2011) Kennedy space center's command and control system: toasters to rocket ship. In: NASA Project management challenge, Long Beach, USA, Feb. 9–10
6. Clarke SC (2011) Project Orion abort flight test: overview of avionics and electrical ground support equipment. In: 49th AIAA aerospace sciences meeting: orion pad abort 1 flight test, Orlando, USA, January 4–7
7. Ren JT, Cai YW, Tong J (2008) Research on establishment annd evaluation of test and launch procedure of launch vehicle. Missiles Space Vehicles (6):4–7,12 (in Chinese)
8. Zhou JP (2013) Space rendezvous and docking technology. National Defense Industry Press, Beijing, 2 (in Chinese)
9. Song ZY (2014) High reliability design for launch vehicle control system. China Aerospace Publishing House, Beijing (in Chinese)
10. Mark M, Shelly E (1992) Advances in ground operations for the next generation space launch vehicle programs. In: Proceedings of the 3rd international conference on engineering, construction, and operations in space, Denver, CO, USA, May 31–Jun 4

11. Giangarra P (2011) Bruce semple. A COTS (Commercial off the Shelf) based approach to plug-and-play launch control system. In: 4th IEEE international conference on space mission challenges for information technology, Palo Alto, USA, August 2–4
12. Nail W, Scheuermann P, Witcher K The use of programmable logic controllers (PLC) for rocket engine componet testing. AIAA, AIAA Paper 91–2403
13. Song ZY (2012) The design of PLC based on hardware logic solver and its application in test and launch control system. Aerospace Control 30(5):78–82 (in Chinese)
14. Wan WM, Zhang X (2002) A study of urgent-off while carrier rocket launching. J Astronautics 23(4):52–56 (in Chinese)
15. GJB 2991A-2008 General specification for B time code interface terminal, 2008. (National military standard, in Chinese)
16. Chen YS (2014) Design and implementation of FPGA-based IRIG-B (DC) decoder. Radio Communications Technol 40(1):93–96 (in Chinese)
17. Nie H (1997) IRIG-B(AC) code demodulation technology analysis. J Spacecraft TT&C Technol (1):1–5 (in Chinese)
18. Yu LF, Zhang L (2008) Network design of test launch and control system for launch vehicle. Aerospace Control 26(4):68–74 (in Chinese)
19. Zhang CG, Chen MK, Wang G (2008) Ground real time network system of launch vehicle. Missiles Space Vehicles (4):19–21 (in Chinese)
20. Lu HR,Lv M (2010) A highly reliable and reconfigurable monitoring and control network system. Missiles Space Vehicles (1):7–9, 51 (in Chinese)
21. Zhang CG, Yi H, Li C et al (2013) Research on technology of assembly and testing for launch vehicle. Missiles Space Vehicles 2:71–73 (in Chinese)
22. Mohoit J, Mayorga A (2006) An auto-configuration system for the GMSEC architecture and API. In: Space mission challenges for information technology 2006, Pasadena, USA, July 17–20
23. Smith D, Bristow J, Crouse P (2007) Reducing development and operations costs using NASA's "GMSEC" systems architecture. In: 7th international symposium reducing the costs of spacecraft ground systems and operations (RCSGSO), Moscow, Russia, June 11–15
24. Pomales WT Software Fault Tolerance: a Tutorial. NASA, NASA/TM-2000-210616
25. Shi Z, Zhou XL, Wang W (1996) Software reliability design guide. Q Reliab (4):38–42 (in Chinese)
26. Hill JV (1988) The development of high reliability software-RR&A's experience for safety critical systems. In: Proceeding of 2nd IEE/BCS Conference on Software Engineering, Liverpool, England, July 11–15
27. Bologna S, Leveson NG (1986) Special issue on reliability and safety in real time process control. IEEE Trans Softw Eng SE-12(9):877–896
28. Ground operations aerospace language, NASA, NASA-TM-X-69541
29. Desjardins R, Hall G, Mcguire P et al GSFC system test and operation language (STOL) functional requirements and language description. NASA, NASA-TM-79541
30. Wilhelm L (1985) KSC space station operations language (SSOL). In: Computer science and data systems technical symposium, Washington, USA, Aug 01
31. Mims TL (2008) Use of spacecraft command language for advanced command and control applications. NASA, KSC-2008-288
32. Baldi A, Elgaard D, Lynenskjold S et al (1994) SCOSII OL: A dedicated language for mission operations. In: 3rd international symposium on space mission operations and ground data systems, Greenbelt, MD, USA, Nov 15–18
33. Ritter G, Pedoto R (2010) The next generation of ground operations command and scripting in C# and Visubal Basic. In: SpaceOps 2010 conference, Huntsville, USA. April 25–30
34. Liu JY (2004) Script description and software design method for test of control system in launch vehicle. Aerospace Control 22(1):73–77 (in Chinese)
35. Aguilar ML, Bonanne KH, Favretto JA et al Review of Ground Systems Development and Operations (GSDO) Tools for Verifying Command and Control Software. NASA,NASA/TM 2014-218278

36. Nidhra S, Dondeti J (2012) Black box and white box testing technique—a literature review. Int J Adv Comput Sci Appl 2(2):29–50
37. Kelly JH, Dan SV, John JC et al A practicaltutorial on modified condition/decision coverage. NASA, NASA/TM-2001-210876
38. Khan ME, Khan F (2012) A comparative study of white box, black box and grey box testing techniques. Int J Adv Comput Sci Appl 3(6):12–15
39. Misra S (2003) Evaluation four white-box yest coverage methodologies. In: Canadian conference on electrical and computer engineering (CCECE), Montreal, Quebec, Canada, May 4–7
40. Tahat LH, Vaysburg B, Korel B et al (2001) Requirement-based automated black-box test generation. In: 25th annual international computer software and applications conference (COMPSAC), Chicago, IL, USA, Oct. 8–12
41. Kruse PM, Wegener J, Wappler S (2009) A highly configurable test system for evolutionary black-box testing of embedded systems. In: proceedings of the 11th annual conference on genetic and evolutionary computation (GECCO), Montreal, Québec, Canada, July 8–12
42. Rajan A (2006) Coverage metrics to measure adequacy of black-box test suites. In: 21st IEEE international conference on automated software engineering (ASE'06), Tokyo, Japan, Sept. 18–22

Chapter 7
Responsive Test and Launch Control

Simplifying test and control operation, reducing launch preparation time, and limiting technical support personnel required on site, are apparently contradictory to improving test coverage, reducing operation differences between ground test and onboard flight, and improving testing facility versatility, but these are all important issues for TLCS [1]. Traditional TLCS has several shortcomings, as follows

- Many technical support personnel are required on site, with many professionals involved to ensure success, particularly for data analysis. This significantly increases workforce costs. Although some solutions have been proposed, such as relaying test data to the design department far from the launch site, analysis still requires more personnel. An alternate approach could be to automate data analysis, but this is often limited to few fixed test cases, so the personnel savings come at the expense of test coverage.
- Many on-site operations are manual. Although the control system manual operation is significantly less than LV assembly, every manual operation has flow-on effects, e.g., connecting or disconnecting a plug requires staff and raises the chance of incorrect operation. To reduce the latter risk, further staff are required to check and report the outcomes. In addition, the launch zone, in particular, has many hidden dangers for manual operations, affecting product and personal safety.
- Testing efficiency is low. Current test schemes have been in use for decades, and the three fixed types of general check no longer cover various flight scenarios. However, if new test cases are introduced, state conversion and data analysis will require more trained operators. Although simulation tests can change test states, in most situations the systems tested are not in their final state. Fixed test cases in the general check, particularly open-loop tests, reduce test validity and fail to meet the requirements for closed loop guidance, e.g., iterative guidance.
- A lot of launch preparation work.

Due to these shortcomings, as well as prelaunch element generation and testing, launch operation is extremely complicated.

Z. Song, *The Test and Launch Control Technology for Launch Vehicles*,
https://doi.org/10.1007/978-981-10-8712-7_7

On the one hand, test coverage and data analysis requirements are steadily increasing to meet improved quality control demands, with consequential increases in the amount of data analyzed and its complexity, all of which increases time and human resource overheads, as well raising training levels. On the other hand, responsive, reliable, and economical access to and from space has become an important indicator of rocket competitiveness. These two requirements are contradictory.

The purpose of data analysis is to ensure that the rocket is in normal working condition before takeoff, and fault diagnosis is critical. Aerospace sectors in all countries have studied automated expert systems to identify and diagnose failure causes, with the intention of reducing technical support personnel while retaining or improving accuracy. However, applications have been less than ideal due to an incomplete and inaccurate KB.

The requirement for long-distance launch support systems has also become critical. Test data can be relayed to the design and production department through the network, and expert systems have again been investigated. Expert systems for fault diagnosis were the earliest practical applications for intelligent technology, and may play a role when the LV information level is low and test data is difficult to obtain. However, technology developments have removed some weaknesses, and fault diagnosis has extended beyond expert systems, allowing further integration and application of new technologies.

This chapter introduces the fast test and launch scheme, then discusses the development of fault diagnosis technology and its application to LVs, and proposes a data-driven model-based automatic analysis system, which takes advantage of recent BIT developments and current high-speed communication between OBS and GS. The proposed method is particularly suitable for pre-launch checks, with significantly improved adaptability, accuracy, and convenience compared with other fault diagnosis technologies. Finally, we propose a closed-loop integrated test method that combines the advantages of traditional integration and simulation testing, and further enhances system level test effectiveness.

7.1 Fast Test and Launch

Continuing the original design routne will not achieve significant efficiency improvements. Therefore, we review the basis of the problem, without limiting the focus to the test system itself. Gradual improvements have struggled to enhance TLCS performance, and OBS and GS requirements must be considered from an overall perspective. Chinese launch system studies emphasize that smart, comprehensive, and convenient are the urgent TLCS needs.

- Smart means to reduce professional support requirements by intelligent data analysis and interpretation, which must adapt to the different test cases.

- Comprehensive means to improve test coverage and consistency between test and flight using closed-loop techniques rather than open-loop testing, paying equal attention to functional and performance tests.
- Convenient means to simplify on-site operation with the ultimate aim of unmanned operation in the launch zone, ensuring launch elements can be automatically generated onboard in real-time, improving LV adaptability and usability.

7.1.1 Simplify Launch Operation

Figure 7.1 shows the current TLCS workflow (process A) [2], based on launch processes of existing LVs, and focusing on operations in the launch zone from propellant filling to ignition. The details differ slightly for different LVs. The activities marked with underlines are those that require human resources, i.e., manual operations, and hence are relatively time consuming.

1. Pyrotechnics loop resistance test. Usually, external test equipment is required, with many test cables and connections that are then disconnected after testing. The tested system state must then be restored, including pyrotechnics short-circuit protection.
2. Battery installation. Current batteries are not designed for reuse and have limited operation cycle life. They are seldom used in launch site tests, and are usually installed just before launch.
3. Status check. The first two operations, among other, change electrical connection states frequently, hence status check and confirmation constitute a significant workload before and after each test phase.
4. Launch element preparation. The elements are transformed by ground software, and linked with other parts of the flight software. Software regeneration must be tested (linked and compiled), then reviewed before hand-over.

Various improvement measures are proposed, respectively, [3] to address these issues, as shown in Fig. 7.1 (process B).

1. Onboard pyrotechnics special test equipment (PSTE) is employed for loop resistance tests, removing the requirement for external test equipment. The PSTE can automatically implement testing, short-circuit protection, protection relief, etc. This not only shortens the testing time, but also reduces plug operations, dramatically reducing status check items.
2. Reusable batteries are employed for test and flight. Therefore, before the LV is transferred to the launch zone, the battery is installed, connected, and fully charged.

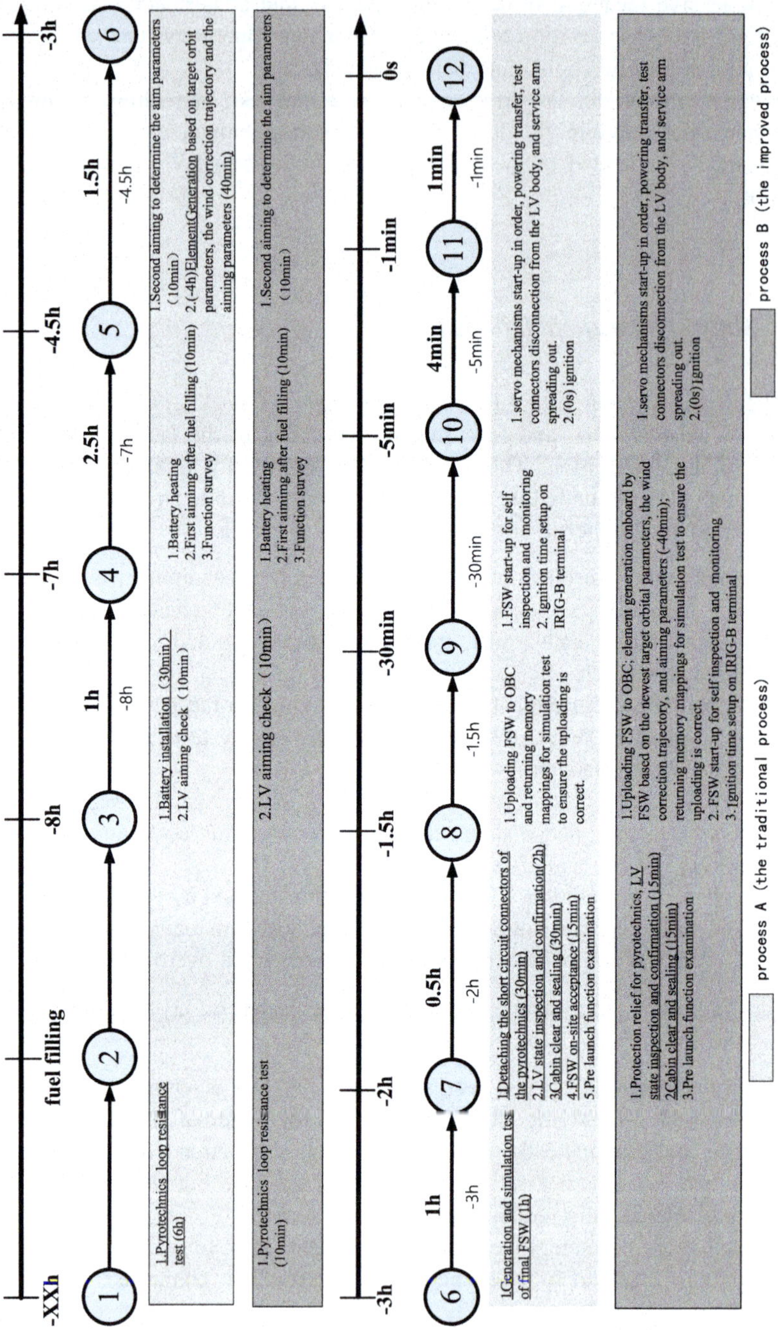

Fig. 7.1 Launching workflow

3. Measures 1 and 2 means LV state after testing is unchanged, greatly reducing the required check items, and at least significantly reducing, if not completely removing, manual operations in the launch tower and launch zone.
4. Closed-loop guidance methods greatly reduce the number of launch elements, and hence code volume for element conversion. Thus, the code can be integrated into the FSW, and previously ground-based element generation can be performed onboard. The code can be tested and validated in advance along with other FSW functions, so special prelaunch tests for elements and final flight software can be omitted.

Comparing processes A and B of Fig. 7.1, approximately 350 min can be saved before fuel filling, and 295 min for the workflow commencing 8 h before launch. These are significant savings, and a major step toward unmanned launch zone operation.

7.1.1.1 Pyrotechnic Checking, Protection, and Relief

To avoid accidental pyrotechnics ignition during testing, the pyrotechnic ends must be short-circuited, usually by connecting short-circuit plugs, which must be manually removed before launch. This is a somewhat risky launch tower operation. Once the pyrotechnics are reconnected to the LV electrical system, the pyrotechnics loop resistance is checked for short or open circuits, or other situations outside the design requirements. This test is quite time consuming.

Considering the risk and time required, there was strong interest in automating this process. Figure 7.2 shows automatic short-circuit protection, and R_h represents the pyrotechnics. The two pyrotechnics ends are connected to a resistor box using relay contacts. When the contacts are closed, the circuit is shorted (safe), and the

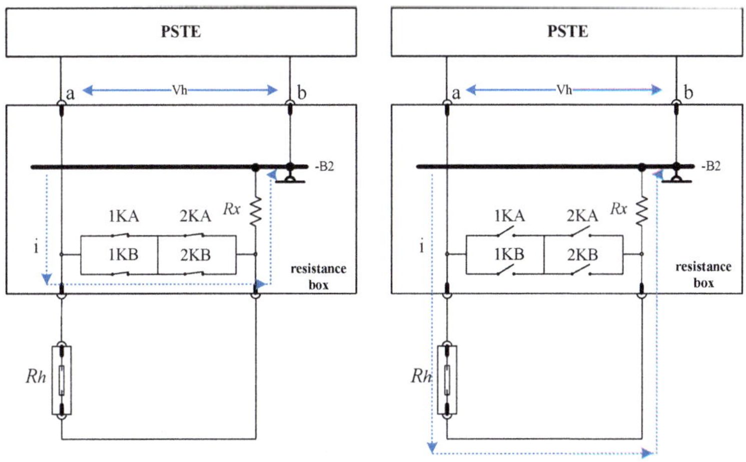

Fig. 7.2 Pyrotechnics short-circuit protection schematic diagram

contacts may be separated, i.e., relieving that protection, at any time before launch. The contacts adopt the parallel–serial configuration to ensure reliable disconnection before launch in case of the single degree of failure, guaranteeing normal explosion during flight. When the short-circuit is required during testing, the state can be verified as discussed in the following section, i.e., by measuring loop resistance.

The relay control scheme is shown in Fig. 7.3. Symbols 1 and 2 K represent two magnetic latching relays, A and B represent the two sets of contacts, and X and Y represent the pre- and post-excitation coils, respectively. There are two exciting coils in the magnetic latching relay, and their excitation state is indicated by test relay TK. The excitation coils are connected in parallel with the close contacts of the relay (JS1–JS4) to form the locking circuit. When excitation is required, the locking circuit is relieved first, which effectively avoids any interference signal accidentally triggering the relay.

An onboard PSTE is configured for automatic testing, replacing the external manual tester, as shown in Fig. 7.4. A similar test scheme has already been implemented for some other applications [4]. The PSTE applies a small constant

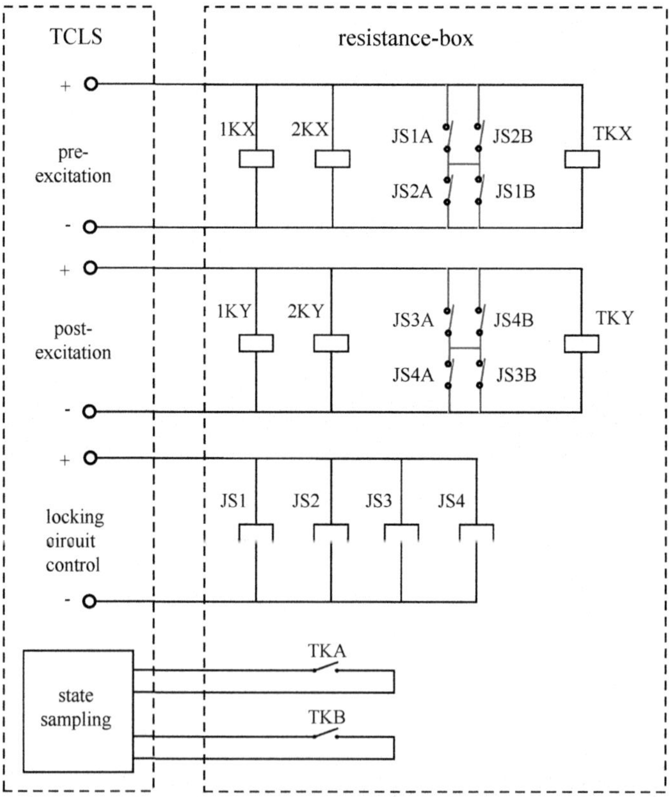

Fig. 7.3 Control and detection of the pyrotechnics short-circuit protection

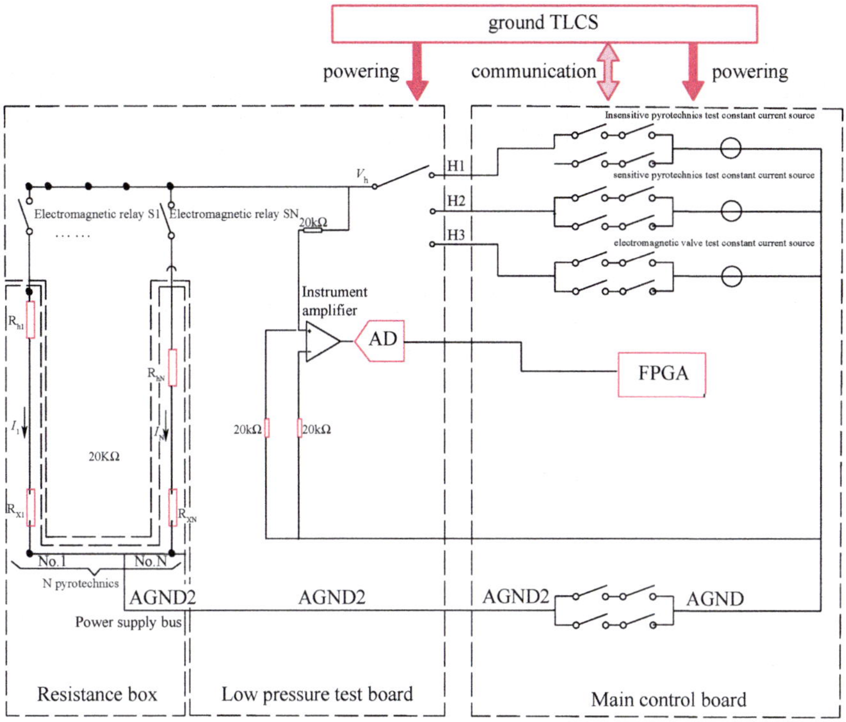

Fig. 7.4 Loop resistance test schematic diagram

current, I_1, to the measured circuit, which flows through the pyrotechnics (R_h) and current limiter (R_x) resistances. Given the measured voltage, V_h, circuit resistance can be obtained by calculating V_h/I_1. When the pyrotechnics are short-circuited (protected), pyrotechnics resistance $R_h \approx 0\ \Omega$, and the calculated resistance is just R_x; when the protection is relieved, the calculated resistance is $R_h + R_x$, and subtracting R_x provides R_h.

The workflow for short-circuit protection and relief is shown in Fig. 7.5. Function A ensures the short-circuit status when the pyrotechnics are connected to avoid accidental ignition. Function B ensures protection relief before takeoff. With this design, protection relief can be delayed to a convenient time before takeoff, further enhancing personnel and product safety before launch.

7.1.1.2 Reusable Battery Employed for Ground Testing

Pyrotechnics or electromagnetic valve loads are replaced by resistive loads for ground testing, called pyrotechnics and electromagnetic valve equivalents. The resistance is generally 1–2 kΩ, and the current consumption is low. Hence, these

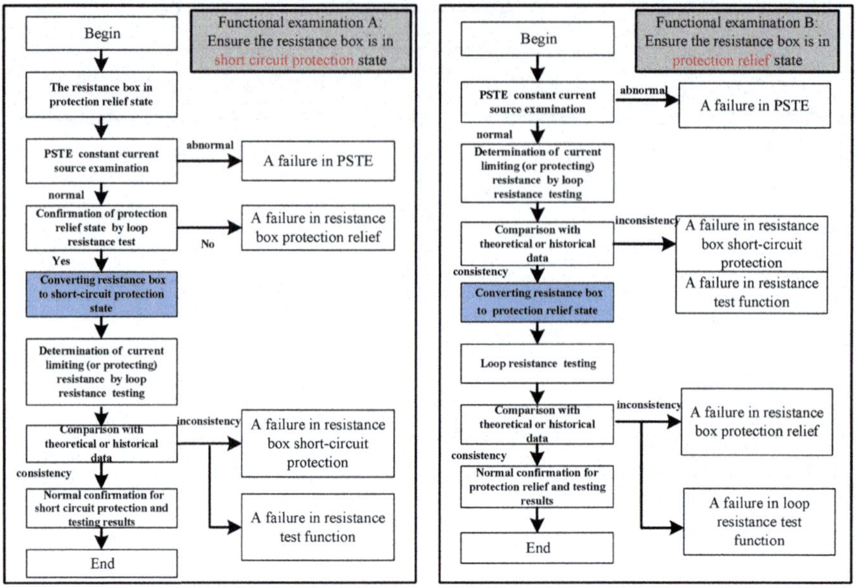

Fig. 7.5 The workflow for short-circuit protection and relief

equivalents can be powered by reusable batteries rather than the ground power supply during testing, not only due to low current consumption, but also to take advantage of prominent reusable battery advantages such as cycle times and charging speed. The batteries are installed onboard to maintain the powering connection consistent with flight, and they can also replace other ground lightly loaded PSs. The benefits are as follows.

1. Ground powering cables are no longer required for battery-powered equipment. This is particularly useful to limit the number of pullout plugs and cables on the LV tail to connect to the ground PS.
2. Test status is closer to the real situation, with no simulated power cable required (Sect. 6.2.2.1). If the reusable batteries are mature and suitable for flight control, the powering status during ground testing will be identical to flight status.
3. Improved power reliability. Front-end equipment in the launch pad bear impact and other loads during launch, the physical space is very limited. Using onboard batteries will help reduce front-end PSs, and simplify ground facilities.
4. More suitable for onboard distributed control systems, where each stage has its own separate in-flight PS system. Using onboard batteries, there is no requirement to configure ground PSs according to every stage PS system, significantly simplifying OBS and GS powering connections.

The scheme has some preconditions, e.g., if the emergency shutdown is controlled by GS with large current, then ground power supplies will still be required.

7.1.1.3 Prelaunch Element Preparation

There is sufficient time for satellite launch to prepare and test launch elements, because satellite target orbits can be determined very early. The control system can update related parameters more than 20 days before launch, and all elements can be prepared in advance. However, rendezvous and docking missions generally cannot set target orbits early, and require orbit parameters to be tested near launch time to avoid large forecasting errors. Testing close to launch time means significantly less available time for element preparation. Rendezvous and docking mission orbits in China are generally determined 4 h before liftoff, so element preparation must be completed within this timeframe, including testing, flight software regeneration, review, archiving, and upload to the OBC. This increases prelaunch operation urgency.

If preparation is done automatically by flight software, ground workload can be greatly reduced and the whole launch process simplified. Iterative guidance helps, because only five orbital launch elements are required, compared to the complete standard trajectory parameters for perturbation guidance. Thus, element generation is greatly simplified, as shown in Fig. 7.6.

The conversion algorithm in generation is simple, and could be integrated in the FSW with almost no effect on software complexity and reliability. Conversion is achieved quickly and automatically by the onboard process before liftoff, and does not affect OBC runtime in flight. However, this is not the preferred approach for perturbation guidance element generation.

On the other hand, orbit determination constraints can also be relaxed with the help of communication technology. Orbit parameters can be measured and relayed to the launcher OBC in real time through the launch site C3I, TLCS network, and GS and OBS communication, and this process can continue right up to power

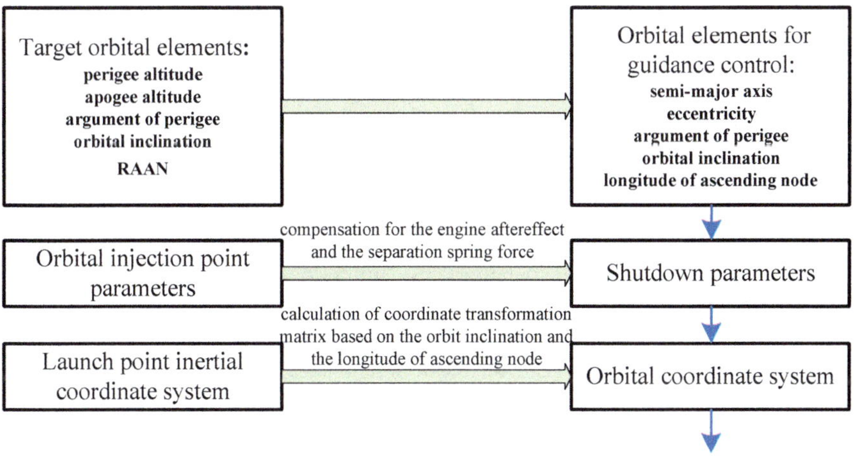

Fig. 7.6 Element generation for iterative guidance

switch before liftoff. The parameters will be more accurate than any forecast 4 h prior to it. Even after liftoff, the parameters can be corrected by reliable wireless transmission, such as Beidou SMS, before iterative guidance is introduced. These all contribute to improved orbit accuracy and LV adaptability.

7.1.2 *Improve Test Coverage and Effectiveness*

Simplifying the operation cannot be at the expense of simplifying test cases or requirements. On the contrary, test coverage and effectiveness should be improved, with every test opportunity seized to examine and evaluate the system as accurately and comprehensively as possible. Testing is divided into two areas as discussed in the following sections.

7.1.2.1 Subsystem Testing

Subsystem testing comprises comprehensive static examinations. Test equipment capacity and real-time constraints mean subsystem tests are often performed serially in sequence. However, this is time consuming, and only some test items can be sampled before liftoff. The following improvements can be considered.

1. Combining onboard BIT functions to realize continuous parallel testing and merge test items to reduce test status conversion and shorten test time. These significantly shorter tests can be performed any time right before liftoff, not just as sampling tests.
2. Use bus interception and high-speed measurement buses (HMBs) [3], to achieve continuous sampling and monitoring of all OBS input and output signals (see Sect. 7.2.3.2 for details).

7.1.2.2 General Check

The general check is a comprehensive simulation of flight processes. The traditional solution employs open-loop tests with static-state IMU output as the system input. This is unsuitable to test closed-loop guidance. The general check also has different powering configurations, due to battery life limitations, which cause many state conversions. This is somewhat complicated and varies significantly from real flight states. Reusable batteries (Sect. 7.1.1.2) and closed-loop testing strategies (Sect. 7.4) [3, 5] can compensate for these deficiencies, as shown in Table 7.1.

Table 7.1 Comparison of different test states

	Test state			Traditional LV	Developing LV	Future LV	Supplement
1	Powering state	State I[a]: OBS and GS connectors are detached following the planned time sequence during flight simulation, sometimes a connection cable for power-off control is remained	Powered by simulated power cable[b]	✓	–	–	
			Powered by batteries	✓	✓	✓	Due to the life cycle limitation, the battery used for the test may not be used for the flight. The application of reusable battery can maintain the state consistency between test and flight
		State II[a]: all connectors between OBS and GS remain connected during flight simulation	Powered by ground PS	✓	–	–	The state II test will be canceled in the future and replaced by state I test, which is closer to the real flight state
		State III[a] The pre- launch process is strictly simulated, and the emergency shutdown is triggered after ignition	Powered by simulated power cable[b]	✓	–	–	
			Powered by batteries	✓	✓	✓	The battery used for this test may be used for flight

(continued)

Table 7.1 (continued)

	Test state		Traditional LV	Developing LV	Future LV	Supplement
2	Cutoff state	Various test cases are set to check exhaustion, guidance, timing, low overload, and other backup cutoff modes	✓	✓	✓	
3	Closed loop check	The OBC sends a control command to drive servos to swing the engines, causing the LV motion that is sensed by the IMU, then the FSW samples the current IMU outputs and generates updated control command. In this way, a closed loop control is formed	–	–	✓	The perturbance guidance is open loop control, and the iterative guidance flight simulation with data replacement method[c] is not closed loop also. Some new closed-loop flight simulation approaches should be adopted in the future LV
4	Deviation (interference) state	Various system parameters deviations are set to check control system adaptability	–	–	✓	The open loop test scheme is unsuitable for imposing deviations or interferences. With closed-loop test approach, the LV motion simulation is integrated into the system testing loop, and the deviations setting is more convenient to enrich test cases

[a]See Sect. 4.2.1.2
[b]See Sect. 6.2.2.1
[c]See Sect. 4.2.1.2

7.1.3 Reducing Technical Support Personnel

Launch site technicians can be divided into three groups.

1. Product support personnel. They deal with on-site failures. Product maturity means this staff level is decreasing.
2. Test personnel or operators. They implement system test and launch preparation, launch control, etc. The number of required operators is decreasing with increasing automation. Onboard operations are greatly reduced when onboard BIT is adopted.
3. Data preparation and analysis.
   Personnel involved in data preparation are primarily for launch element generation. Automatic FSW conversion means full-time personnel can be significantly reduced.
   However, data analysis requires professional knowledge. Although the work has become more convenient, more detailed analysis requirements have also increased technical background requirements, particularly for data coupling or relevance analysis. Indeed, sometimes even only the original designers can solve some issues. On the other hand, manual data analysis is time consuming, and automatic analysis with fixed thresholds are difficult to implement for time-varying parameters. Although some progress had been made, such as feature points-based automatic analysis for time-varying parameters, this approach introduces false negative risks. This method is only adopted for fixed test cases and when test cases vary, the feature points, which are determined and selected manually, also vary, i.e., analysis criteria for the feature points must be prepared in advance.
   Intelligent technology will reduce professional data analyst workloads, potentially significantly. This aspect is discussed in Sect. 7.3.

Thus, three improvement aspects can be realized in the TLCS: simplifying launch operations, improving test coverage and effectiveness, and reducing on-site technical support personnel. Professional requirements for operators and the workforce can also be significantly reduced, possibly to the point of removing all manual operation in the launch tower and launch zone. This has great benefit for operator and product safety, while ensuring apparently seemingly contradictory requirements are met, e.g. test coverage and validity, and simplified operation.

7.2 Pre-launch Diagnosis

Classification of fault diagnosis technologies varies [6–9]; this paper presents launcher-related research achievements by categories: threshold based judgment, knowledge-based expert systems, model-based fault diagnosis, self-learning-based reasoning mechanisms, information fusion-based fault diagnosis, and self or automatic diagnosis. Some methods are too complex for pre-launch checks.

7.2.1 Expert System

Expert systems are an important and active field of artificial intelligence research and application. First proposed in the 1960s, they developed rapidly in the 1970s–1980s, and have been widely adopted in industrial and national defense domains. Expert systems were introduced into LV control systems in the 1980s, mainly for fault diagnosis, etc. This naturally attracted the attention of TLCS designers.

An expert system is an intelligent computer program [10] that contains massive (human) expert knowledge and experience, designed to deal with high-level problems. The KB, DB, and reasoning machine comprise the three elements of an expert system.

Important relevant artificial intelligence and knowledge engineering concepts include knowledge representation, acquisition, and reasoning. Knowledge representation refers to knowledge expressions in the computer, which include state space, problem reduction, predicate logic, semantic network, frame, script, production rules, etc.

Production rules are widely used where the KB is equivalent to a rule base. A production rule basic form is usually expressed as

IF <precondition> THEN <conclusion/action>,

where the IF statement is called the precondition, or the left of the production rule; and the THEN statement is regarded as the result, action, or the right of the production rule; e.g.,

IF <receive booster propellant exhaust signal> THEN <turn off booster engine in T seconds>

In general, the precondition applies the logical "AND" for many facts. The reasoning machine first matches the precondition in the current database; if satisfied, the rule can be applied, and the conclusion or relevant action obtained. If more than one rule matches the current database, conflict resolution decides which rule to apply first. Production rules are based on deductive reasoning, rule preconditions are matched repeatedly with facts in the database, and conflicts are solved continuously, forming a reasoning chain to achieve the final conclusion. Production systems based on knowledge expression rules have the advantages of being intuitive, clear, flexible, convenient, and modular, and have been widely adopted.

The reasoning process can be divided into forward, backward, and bidirectional reasoning, depending on the search direction. Forward reasoning is also called fact driven, data driven, or forward chaining reasoning, and starts from the precondition in the database to query all rules and generate the final conclusion. For complex reasoning, the conclusion formed during the processing can also be added to the database as a new fact.

Backward reasoning is also called goal driven, hypothesis driven, or backward chaining reasoning, and starts from the target or conclusion to find rules that matched the conclusion. It reverses links to all relevant rules until a precondition that supports the expectation conclusion or a contradictory conclusion is found.

Table 7.2 Comparison of forward and backward reasoning

Item	Forward reasoning	Backward reasoning
Driving mode	Data driven	Target driven
Reasoning method	Forward deducing a conclusion from a set of data	Backward reasoning to verify a solution from a possible solution
Startup mode	Start from an event	Start by inquiring a question about the target state
Reasoning direction	From the bottom to the top	From the top to the bottom
Transparency Degree	Unable to explain the reasoning process	Able to explain the reasoning process
Typical system	CLIPS	PROLOG

Table 7.2 compares forward and backward reasoning. Forward reasoning is mainly used for questions where the initial data is known, but no reasoning target is provided; or where the solution space is very large, such as monitoring and prediction. Backward reasoning is mainly used for situations where the conclusion is unitary, or the conclusion is known but needs to be verified or causes identified, such as failure diagnosis.

There are three ways to acquire knowledge.

1. Compiled by the knowledge engineer. The knowledge engineer talks with domain experts then summarizes the acquired knowledge, compiles the software, and debugs repeatedly. This is a relatively long process.
2. From the KB editor. The engineer designs the KB editor, and knowledge is obtained through dialog with a domain expert using the editor.
3. Employ machine learning to acquire knowledge, e.g., training with the neural networks.

7.2.2 Common Diagnostic Mechanisms

Classification of this section is not absolute, e.g., some GS diagnosis techniques may also be applied to onboard self-diagnosis; for model-based diagnosis, the model can be regarded as the KB; and self-learning mechanisms form a diagnostic model. Therefore, the classification here is distinguished broadly by the technological development pathway.

7.2.2.1 Threshold-Based Judgment

Threshold-based judgment is the earliest and most commonly used method. When the signal exceeds the threshold, it shall be considered a failure. This is mainly applied to static tests, slowly varying, or relatively fixed test cases.

If the signal varies continuously, it is difficult to set a time-varying and synchronous threshold. Signal features or characteristics based analysis are introduced for some situations (Sect. 7.1.3), i.e., only limited data or particular periods are checked, these time points are usually when attitude control network parameters change, e.g., engine shutdown, stage separation, engine ignition, etc. Moreover, test criteria should be prepared in advance for different test cases.

Fault propagation should also be considered for closed-loop feedback control systems. Changing input conditions cause varied outputs, which change the system state, and hence the input conditions. The fault is propagated, and each part of the closed-loop system varies from the original state, which disturbs the fault diagnosis. Therefore, signal processing, such as frequency domain and time–frequency domain analysis has been developed.

Signal processing transforms original time domain signals into other quantities, and seeks judgment criterion for these transformed quantities. Commonly employed transformations include empirical mode decomposition (EMD) and information entropy-based technology [11]. EMD analyzes the energy and frequency using changes of the signal's internal timescales, expands the signal into a number of intrinsic mode functions (IMFs). IMF instantaneous frequency and amplitude are obtained from the Hilbert transform, and then transformed into information entropy as the failure detection features.

Threshold-based judgment is not regarded as intelligent diagnosis, but all the signal processing methods can be used for intelligent analysis.

7.2.2.2 Rule-Based Reasoning

When the threshold-based method does not meet the requirements, the expert system is considered earliest as a way of artificial intelligence processing [12, 13]. The Ref. [12] introduces that Beijing Aerospace Automatic Control Institute collaborated with Beijing Institute of Technology, and they designed a fault diagnosis expert system for one LM launcher. The system has more than 2000 rules in KB related to 21,000 knowledge points, and it is the first practical application for Chinese LVs.

Rule-based reasoning is widely used in expert systems. Consider the example of a judgment for boost shutdown to illustrate the principle. To make full use of LV carrying capacity, booster shutdown control prefers exhausted mode, i.e., fuel is utilized as much as possible. Hence theoretical boost shutdown time may be very close to that of the first stage, or even lag behind it. In this situation, the booster shuts down first, and then the core first stage shuts down. Suppose the exhaust signal is received shortly after takeoff, which should be determined as a false signal. Therefore, shutdown permission constraints are applied, i.e., before the permitted setting time, all exhaust signals are ignored. On the other hand, if the exhaust signal has not been valid for a very long time, it may be caused by failure of exhaust sensors or transmission links, so timed shutdown control is also set.

If forward reasoning is used to determine whether the booster shutdown signal is correct, the following rules are stored in the KB.

R1: (shutdown permission control)

IF [flight time is less than permitted setting time]
THEN [boost shutdown control is not allowed]

R2: (exhaustion control)

IF [flight time is greater than permitted setting time] &
[booster has not yet been cut off] &
[exhaustion signal is received]
THEN [shut down boost engines in T seconds]

R3: (backup shutdown control)

IF [flight time is greater than permitted setting time] &
[booster has not yet been cut off] &
[core first stage shutdown signal is received]
THEN [shut down the boost engines immediately]

R4: (timed shutdown control)

IF [flight time is greater than or equal to timed setting] &
[booster has not yet been cut off]
THEN [shut down the boost engines immediately]

Therefore, the state is examined depending on flight time, exhaustion signal, core-stage shutdown signal, to judge the action after receiving the booster shutdown signal, based on the four rules. If one rule matches all measured data, the signal is considered as correct; whereas if all rules do not match the data, a fault exists.

In backward reasoning, the conclusion is assumed correct. Consider the example of the timed shutdown. Flight time is examined whether it exceeds the timed setting, and there was no other exhaustion signal or core-stage shutdown signal previously. If the two preconditions are met, then R4 is confirmed, and the signal is considered correct.

The above rules can only judge whether the booster shutdown signal is correct or not, the failure mode should be deduced from fault tree analysis (FTA) [14–16]. Once the bottom events of the fault tree are sorted, rules can be constructed related to the failure.

For example, consider the fault tree shown in Fig. 7.7. For the top event, T, all minimal cut sets can be obtained using the descending method: $\{X1\}$, $\{X4, X7\}$, $\{X5, X7\}$, $\{X3\}$, $\{X6\}$, $\{X8\}$, and $\{X2\}$. Then the following rules are formed:

R1: IF *X1* THEN *T*
R2: IF *X4* & *X7* THEN *T*
R3: IF *X5* & *X7* THEN *T*
R4: IF *X3* THEN *T*
R5: IF *X6* THEN *T*
R6: IF *X8* THEN *T*
R7:. IF *X2* THEN *T*

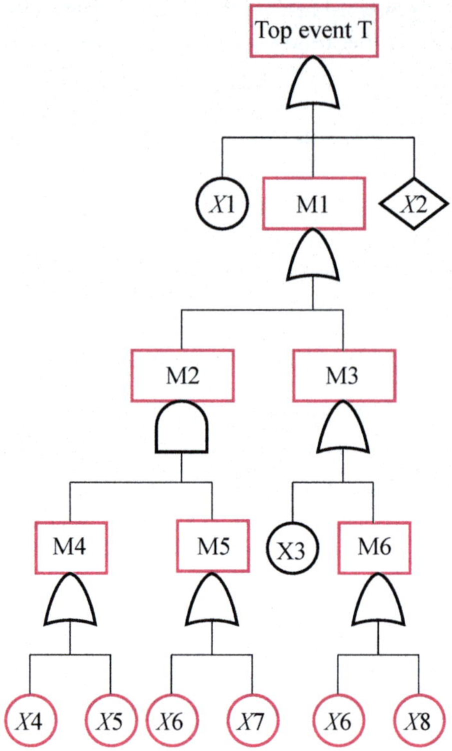

Fig. 7.7 Fault tree example

This reasoning mechanism matches the fault with failure modes, and the fault condition is judged based on test data analysis. Therefore, this method is often employed for failure diagnosis. When fault T occurs, seven possible causes are listed for the operators. Although $\{X1\}$(R1), $\{X3\}$(R4), $\{X6\}$(R5), $\{X8\}$(R6), and $\{X2\}$(R7) are possible failure modes, if these conditions cannot be verified by forward or backward reasoning, then the failure causes are must be R2 and R3. If preconditions $X4$ and $X7$ are established from the test data analysis, then at least one rule, R2, maybe the failure mode.

7.2.2.3 Case-Based Reasoning

Another method of failure location is pattern matching to previous cases [17–21], or case-based reasoning (CBR), which was applied to some early expert systems. However, since failure cases are rare, this method often needs to be combined with other reasoning mechanisms, such as model-based reasoning. It is particularly useful to find previously occurred failure modes.

Consider a typical CBR application case [21]. History is an important part of the KB, and its data structure should facilitate storage and retrieval with the domain value of the data structure referenced in the hypertext document through the internal pointer.

Text matching algorithms play a key role in CBR. Semantics are usually neglected to simplify searches, since they could greatly increase programming complexity. The matching algorithm is totally different for different languages, but modern Internet search engines make matching more convenient. The algorithm should be robust enough that minor spelling errors do not affect search results.

User interaction is another important factor. The system should be easy to browse or study descriptions, failure causes, treatment, and design improvement of former cases.

7.2.2.4 Model-Based Reasoning

Model-based fault diagnosis methods can be adopted where rules and cases cannot adapt to failure reasoning for complex systems. This approach covers a wide range, and models are sometimes seen as a form of knowledge. Description of the system under test can be a physical model, such as dynamics model differential equations, electrical system circuit diagram; or a logical model, such as flight software. Hierarchical models are suitable for large systems. In general, models can be broadly divided into the following categories.

1. Mathematical simulation or design models
   These can be regarded as predication models. When model results are inconsistent with measured values, a failure exists in the system. Knowledge-based automatic test engineer (KATE) [22, 23] is a typical representative of this type of method. KATE simulates the nominal model numerically to detect failures, and simulates the failure model to locate the failure. The simulation results can also be matched with the failure dictionary [24], where the matched item is the failure mode, and this process is very close to that of expert systems.
   Control theory has investigated a variety of robust model-based FDI, including state or parameter estimation [25], extended Kalman filter [26, 27], observer-based diagnosis [28–30], fault-tolerant control using the $H2/H\infty$ norm [31], and convex optimization [32]. This convex optimization considers the difference between predicted and actual flight data as the residual, and the failure feature is the sensitivity or Jacobian matrix depending on possible failure modes. The failure source is determined by solving for maximum likelihood.
   These methods are also known as the analytic redundancy scheme [33], i.e., the model rather than real hardware is considered as the redundant part for failure diagnosis. The difference between the methods mainly reflects adaptation to uncertainty and the degree of tolerance to failure.
   However, method efficiencies in closed-loop feedback control systems are decreased to varying degrees, mostly due to the robust resistance of the

closed-loop system to external disturbance. When the failure is developing, its impact may be compensated by feedback control, resulting in higher false negative rates [34]. Thus, although this area has been widely studied, it is rarely applied to large LVs.
The failure location algorithm can be greatly simplified if BIT information and sensors can be fully utilized to directly obtain measurement signals of suspected fault sources.

2. Functional models
Functional models are a very effective approach to complex system analysis, and provide high-level abstractions and hierarchical designed model architectures. More importantly, some software tools support this model type combined with reliability analysis, forming functional failure model.
The functional simulation model (e.g., GNC or dynamic simulation) can be integrated with reliability analysis tools (e.g., risk analysis, FTA, FMEA, etc.) with one of the important analysis results being FDIR strategies for ground testing. Failure propagation and sensor configuration analyses can also be performed, for a detailed description, see [35].

7.2.2.5 Self-learning-Based Reasoning

Self-learning-based reasoning is also called a data-driven fault diagnosis method. It uses anomaly detection algorithms [36] to identify abnormal situations by analyzing of the data set. Processing of data sets can be divided into two approaches that can both be trained using the neural network algorithm.

- Classify input–output data sets, into normal and abnormal data sets.
- Construct a discriminant model between input and output signals, to judge whether the output signal is normal under the given input.

Classification is a typical neural network application, particularly for linearly separable solution spaces, as shown in Fig. 7.8 for a simple application.

Suppose there are four samples, two are normal, $x^1 = [1, -3]$, $x^2 = [-3, -1]$; and the other two are abnormal, $x^3 = [2, 2]$, $x^4 = [-2, 1]$. Each sample has two elements, representing the simplest single input and single output system. A classification model is established using these four samples. When other input–output pairs are presented, the model judges whether they are normal or not.

Neural networks require a decision function to classify the samples. The asymmetric hard limit function is often suitable for LV applications. If the decision function is $\mathrm{d}(x)$, the perceptron output is

$$\begin{array}{ll} y = 1, & \mathrm{d}(x) \geq 0 \\ y = 0, & \mathrm{d}(x) < 0, \end{array}$$

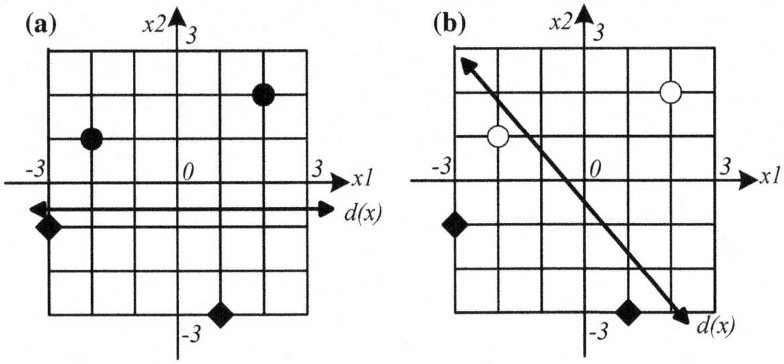

Fig. 7.8 An example of classification

where "1" means normal, and "0" means abnormal. There are many ways to derive decision functions, and this example is relatively simple, so the derivation is omitted, and the decision function is expressed as

$$\mathrm{d}(x) = -3x_2 - 2.$$

Thus, $\mathrm{d}(x^1) = 7$, $\mathrm{d}(x^2) = 1$, $\mathrm{d}(x^3) = -8$, and $\mathrm{d}(x^4) = -5$.

In this example, perceptron output is only related to the system output, x_2, which is coincidental for this case.

Figure 7.8a shows the above classification system is good for classifying normal samples and error samples. Assuming that a new set of input–output pairs, $x^5 = [-2, -2]$, then $\mathrm{d}(x^5) = 4 > 0$. Thus, the perceptron output = 1, and this set of test data is considered normal.

Decision function derivation requires training with normal and abnormal samples, which is called supervised learning. However, if there has been no abnormal situation, i.e., only normal samples are included for training, the training is called unsupervised learning. In this case, the decision function may be different and cause misjudgment. Figure 7.8b shows the decision function using the same derivation where only the first two normal samples were presented for training. The new classification correctly judges the two normal samples (black dots), but also classifies x^4 as normal (white dots), i.e., false negative. Thus, the training sample plays an important role in classification.

The perceptron can be used for failure diagnosis of a single product. Figure 7.9 shows neural networks, which can be applied when multiple products comprise a complex system.

Neural networks have several applications.

1. Combined with failure mode analysis to achieve classification.
   The input layer is the observation data from the object under test, and the output layer is the failure mode or failure bottom event (obtained through FTA) [37]. Application results are intuitive, but abnormal samples are required.

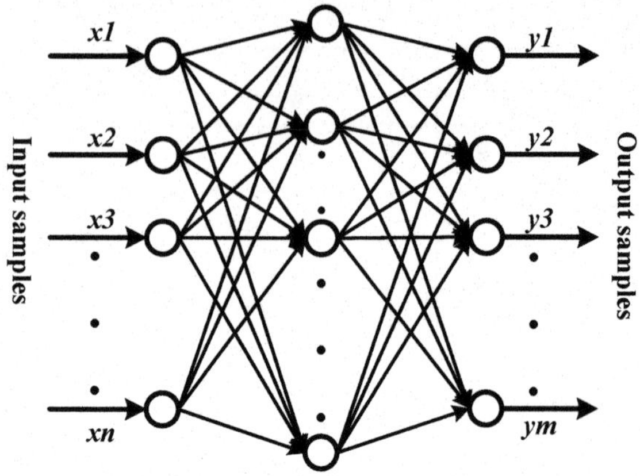

Fig. 7.9 Neural network model

2. Establishing input–output models to identify failures
   Neural networks are widely used in system modeling, and this can also be viewed as model-based reasoning. The input layer is still the observation data of the object under test, but the output layer is the object's output, or a data set that represents system behavior characteristics [38–41]. Neural network output can be trained to match the training samples as much as possible or at least so the error is within a certain range (called the similarity index). A failure is identified as when a new sample exceeds the similarity index.

Self-learning methods include many algorithms, and neural networks trained from input and output information can also be considered as a system model. Therefore, self-learning can also be regarded as model-based reasoning.

Support vector machine methods can also be applied to map the sample space into a high-dimensional or even infinite-dimensional feature space. The failure may be isolated after mapping, while the original sample space cannot be linearly divided into normal or abnormal classifications. Principles and applications of this field are not discussed here.

7.2.2.6 Other Diagnosis Technologies

Information Fusion-Based Fault Diagnosis

If we can increase sensors for key equipment, comprehensive analysis can be made using running state information from these sensors and existing KB. Final evaluation of operation or failure can also be solved by the information fusion. Thus, information fusion has a significant role in failure diagnosis [42].

Popularization and application of distributed control systems mean onboard intelligent equipment have strong BIT and information acquisition ability. Information fusion methods can analyze and synthesize these data to improve synthetic information accuracy. The main methods include the following.

1. Bias theory, i.e., probability statistics-based fault classification [43].
2. Dempster–Shafer (DS) evidence theory, i.e., failure located according to evidence (and consequence) credibility [44].
3. AI combined information fusion, e.g., artificial neural networks based on failure membership degree vector; and artificial immune system that simulates dendritic cells to detect abnormalities [45].
4. Integration of information fusion with other technologies [46].

Although information fusion development is very fast, the algorithms have some limitations, with many engineering problems remaining to be solved. For example, the priori probability in Bayesian method, evidence reliability in DS evidence theory, and membership degree in neural networks, all have a large influence on the algorithm; and the integrated fusion algorithm is very complex.

Autonomous Diagnosis

Most diagnostic technologies introduced require high-speed computing platforms, and are generally deployed on the GS. With increasing signal sampling, it is impractical to transmit all of them to the GS. Therefore, the demand for onboard autonomous diagnosis is increasing.

Conditions for autonomous diagnosis [7, 47] are as follows.

1. Onboard data is transmitted through a network, and transmission speed is increasing, benefiting data collection.
2. OBS processing capacity has greatly expanded.
3. Distributed control systems compensate for low performance of a single onboard computing device.
4. Redundant design is widely adopted for key equipment, and in-flight FDI can also be applied to fault diagnosis before launch.

Autonomous diagnosis mainly adopts the model-based analysis methods introduced above, but signal processing, neural network, and knowledge-based diagnosis, have also been studied. The most common and convenient solutions are redundant information-based processing and BIT, which provide an effective way to implement cross checks, consistency tests, voting mechanisms, etc.

7.2.3 *Example Launcher Health Monitoring System*

7.2.3.1 International Launcher Fault Diagnosis Technology

Almost all LVs have studied fault diagnosis technology. NASA showed that GS FDIR tools can only shorten fault diagnosis time for cases with increased available test data, or help to select important FDIR information [48]. Otherwise, the tools have no effect while increasing system complexity and cost. Fault diagnosis system has gradually developed to an HMS [49, 50], which is defined as the process, method, and technology to prevent failures and reduce failure effects [51], whose core is still fault diagnosis. Thus, in the following discussion, the two concepts of fault diagnosis and health management are no longer strictly distinguished.

The HMS must be planned from the top level [52, 53]. Once the LV is complete, it is too late to consider if there are sufficient test points and signals, and every design change will have potentially significant project costs.

Boeing developed a standardized research and development process for the Delta IV HMS. The process started with the system functional model, identified all potential hazardous conditions, constructed a clue tree for hazard cause, analyzed hazard detection methods, and finally partitioned software functions. A detailed discussion of the process can be found in [54]. A processing and load model was constructed as a feasible detection method to determine HMS hardware processing bandwidth, memory capacity, etc., providing the HMS design requirements. This is basically a system engineering process, and provides a useful reference for LV reliability design and DFT.

For a detailed discussion of other fault diagnosis technologies, see [55–57].

7.2.3.2 Chinese New Generation Launcher Fault Diagnosis Technology

China also undertook expert system research in the early days of the Chinese manned space project [12], but the KB is not adaptable to unknown failures, and was difficult to maintain, with limited measured data available. Meanwhile automatic test data analysis remains the long-term goal, and many threshold-based analysis systems have been developed. In addition, some data browsing platforms can also provide data retrieval. But there is a significant gap between automatic interpretation and intelligent fault diagnosis.

The concept of health management has also affected Chinese LVs. Various solutions, e.g., BIT and bus communication architecture, have been proposed, and health management technical constraints have been studied [58, 59], but have not been deployed yet.

Figure 7.10 shows the new generation Chinese LVs ground TLCS. High-speed digital buses, also called HMBs, were adopted for communication between GS and OBS, so it is feasible to relay significant BIT data from OBS to GS synchronously, which enables model-based fault diagnosis. The GS terminal could use these data and the same model as onboard flight software for calculation, and compare the results to that of the OBS in each control cycle.

The data analysis function block in Fig. 7.10 can reproduce all onboard software processing, including navigation, guidance, control, sequential control, etc. Since almost every onboard controller input and output can be obtained through HMB,

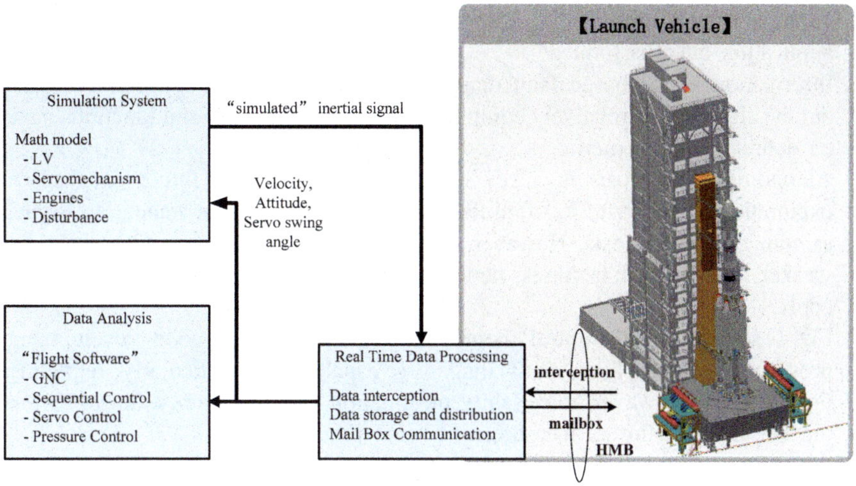

Fig. 7.10 The brief diagram of Chinese new generation LV TLCS

diagnosis uncertainty will be greatly reduced, making fault location very convenient. This will be discussed in detail in subsequent sections.

7.2.4 Launcher Health Monitoring Trends

Although failure diagnosis techniques introduced in Sects. 7.2.1–7.2.3 have shown excellent performance in their applicable areas, some limitations remain.

- Threshold-based judgment relies on signal processing, i.e., it does not require a mathematical model of the object, and hence can extract information features directly. However, it only uses the extracted information, and cannot probe the system structure or internal changes.
- Rule-based reasoning, or knowledge-based expert systems, require more comprehensive and accurate KB. Whether FTA or CBR is applied, the method strongly depends on past experience, and is very effective in finding known faults, but relatively poor in finding unknown or new failure modes.
- Mod-based reasoning can fully utilize deep knowledge of the internal system, and can detect unknown faults, but model uncertainties, external disturbances, and other factors affect judgment accuracy. Robust design of the SUT also increases diagnosis difficulty.
- Self-learning-based reasoning does not need to understand object internal structure, and does not require a KB constructed in advance. It can learn from the samples directly, and map from the representation to the failure source. However, accuracy strongly depends on the number of samples, and particularly

the number of failure samples. Since failure samples are not common in practice, application effect is limited.

- Information fusion-based fault diagnosis integrates many diagnostic methods, but the algorithm is relatively complex, and some parameters and functions have no definite setting criteria.
- Autonomous diagnosis is a key required technology for future deep space exploration spacecraft. It simplifies the TLCS, and is convenient to signal sampling and diagnosis. However, onboard hardware processing capability is weaker than ground facilities, hence many advanced methods are difficult to apply.
- FDI-based advanced control theory and algorithms have good development prospects, and expands fault diagnosis capabilities, particularly regarding uncertainties. However, uncertainty information can be obtained directly from the sensors with further measurement development.

Given these acknowledged limitations, and the requirement for failure detection before launch, the following trends for health monitoring have been identified.

1. Develop from a single technical solution to a variety of solutions.
   Many technical solutions could be adopted to improve fault diagnosis efficiency and reliability and avoid misjudgment. For example, the US Ares rocket uses model-based and learning-based reasoning mechanisms.
2. Model-based analysis will become the major approach.
   With ongoing development of modeling and simulation technology, model-based analysis will become the most common solution for failure diagnosis. Diagnosis algorithms will be greatly simplified if direct measurement data of the suspected fault source can be obtained.
3. Rule-based expert systems will gradually be replaced by other methods
   Building the KB is somewhat inconvenient and the rule itself can also be modeled and integrated into model-based analysis. Therefore, rule-based approaches will be used primarily to complement other methods, or for advice and training.

7.3 Model-Based and Data-Driven Analysis

Test equipment is customized or conforms to certain standards for traditional LV TLCSs, following the design concept that testing is performed by GS equipment, which samples OBS signals. Test technology continues to develop rapidly, and standard test modules and test buses, such as VXI, PXI, etc., are now widely used, which prompts their application to LVs. However, it is not economic for LV test equipment to keep pace with rapid industrial developments, as these standards and practices update too quickly, and it is also difficult to maintain the current state throughout the long LV life cycle due to the lack of consistent spare parts.

BIT technology provides a solution avoiding frequent test module updating, and is more effective for testing. The Japanese solid fuel LV (Epsilon, Sect. 1.3) can check, diagnose, and isolate failures automatically using the OBS, and the go/no-go decision is forwarded to the control center, where only process control runs. However, the limited communication bandwidth between OBS and GS means only test outcomes rather than the original data can be forwarded to the GS, which raises the concern of the LV being out of control.

Another factor considered for improvement is system level test cases. An innovative test scheme must be adopted for new-generation Chinese LVs [60]. We start by looking at how the new LV test requirements differ from previous LVs.

1. Signals connected to ground test equipment have significantly decreased.
   To simplify operation, all connectors installed in the LV outer bulkhead for testing have been removed, leaving only those for launch control located on the LV tail. These connectors are separated during liftoff, and few contacts are assigned for testing.
2. Signals to be measured have significantly increased.
   Control processes for cryogenic engines are more complex than other engines, which increases testing workload.
3. Test coverage is more demanding.
   Each component of the redundant system, not just the overall effect, must be checked.
4. Performance checks are attracting more attention in system level testing.
   Previously, and particularly for the IMU, only the functional examination was practicable during system level testing, which produced many invalid judgments with consequential degraded performance and early failures. Thus, performance check demand is increasing.
5. The number of technicians on the launch site must be reduced to promote efficiency and safety.

To meet the above requirements, we propose a new testing configuration, as shown in Fig. 7.11. Data are sampled by the OBS using BIT, and analysis is performed on the data handling terminal in the control center. If the OBS data can be duplicated in the GS, we can playback onboard processing based on the same algorithms, i.e., models, and compare outcomes with that of the OBS, which will greatly simplify analysis and provide significant FDI benefits. Therefore, an HMB is required between OBS and GS, and data handling terminal processing is triggered as soon as it receives the data, i.e., data-driven analysis. We can apply varying test cases, because the criteria can be formulated concurrently during analysis.

Interfaces between the LV and TLCS, can also be standardized and simplified. Only three types of interfaces: power supply (PWR), discrete input and output (DI/DO), and HMB, are required, and no specialist testing instruments remain in the GS. A communication terminal running an embedded real-time operation system should be installed in the TLCS front-end to receive data from the OBS and distribute them through network to the back-end control center or contractor sites.

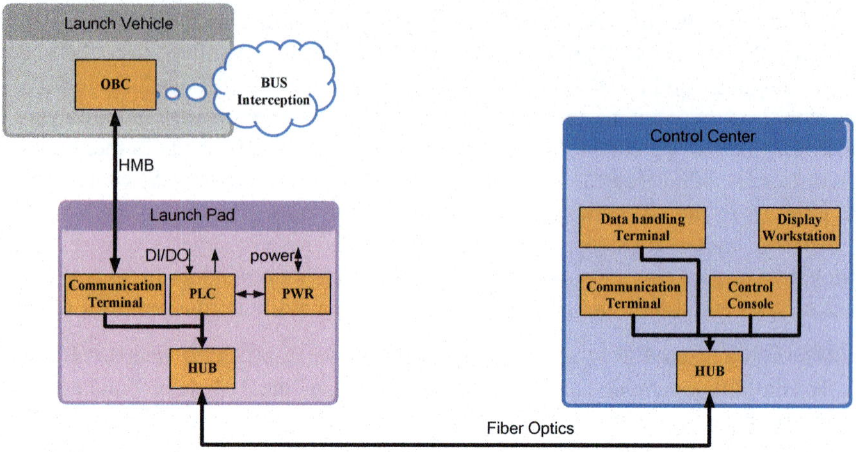

Fig. 7.11 Test configuration based on bus interception

7.3.1 Information Flow Analysis

It is impractical for every LV device to transmit BIT data to the GS. Hence, we need to select an appropriate monitor point to collect the data. Table 7.3 and Fig. 7.12 summarize a typical information flow. There are three types of onboard controllers: OBC; integrated controllers (ICU), e.g., sequential, thrust regulation controllers, etc.; and servo controllers (SCU). The OBC receives signals from inertial devices and exhaust sensors; and transmits attitude control commands to SCUs through the 1553B bus, and cutoff commands to ICUs through the 1553B bus or RS-422 interfaces. ICUs receive exhaust and pressure signals from sensors, and cutoff commands from the OBC; and transmits sequential control (SC) and pressure control (PC) signals. ICUs also perform thrust adjustment. SCUs receive commands from the OBC and drive servomechanisms.

Pressure sensor signals, pressure control commands, sequential control commands, etc., shown in Table 7.3 shadow area, cannot be sensed directly by the OBC. They are encoded in BIT data or telemetry data by the SCUs or ICUs and then sent to the OBC, as shown by the bold arrows in Fig. 7.12. These data, together with important flight control software outputs, would be sent to the telemetry system by the OBC through the 1553B bus.

7.3.2 Bus Interception

We can set the monitor point on the OBC's internal bus. It receives all sensor signals, transmits all commands, and relays BIT or telemetry data to the telemetry

Table 7.3 Characteristic onboard information

Parameter type	Input	Source (interface)	Process device	Output	Actuator (interface)
GNC	Angle, apparent acceleration	IMU (1553B)	OBC	Attitude control command	SCU (1553B)
GNC	Angle, apparent acceleration	IMU (1553B)	OBC	Cutoff command	ICU (1553B, 422)
GNC	Angular rate	Rate gyros (485)	OBC	Attitude control command	SCU (1553B)
GNC	Lateral and normal acceleration	Accelerator assembly (1553B)	OBC	Attitude control command	SCU (1553B)
GNC	Exhaust signal	Sensor (DO)	OBC	Cutoff command	SCU (1553B, 422)
PC	Pressure signal	Sensor (485)	ICU	Pressure control	Valves
SC	Exhaustion signal	Sensor (DO)	ICU	Sequential control	Valves, pyro
SC	Cutoff command	OBC (1553B)	ICU	Sequential control	Valves, pyro.
Attitude control	Attitude control	OBC (1553B)	SCU	Servo drive	Servo mechanism
Thrust adjustment	Start-up command	OBC (1553B)	ICU	Pulse control	Driver and motors

system. All data can be monitored on the bus during processing. We have the following three types of data:

1. FSW inputs and outputs.
2. Controller BIT data.
3. OBC internal data for telemetry (in digital form).

The monitoring technology is called bus interception (BI) [60] to distinguish it from the 1553B bus monitor. It has the following characteristics.

1. Data sampling is automatic, without FSW participation.
2. Data are monitored depending on addresses configured during initialization, e.g., I/O or memory address, etc.
3. The process is triggered in interrupt mode, only when data on a monitored address changes. Transmission is in ping-pong mode through dual FIFO caches to ensure continuity.
4. Hardware interface is based on LVDS. Drivers are required at the transmitting end and adaptive equalizers are added at the receiving end to restore the signals to their original characteristics when sent/received, and avoid attenuation.
5. Software protocol is flexible to non-fixed-length messages marked with source coding and time tags.
6. The receiving terminal has large caches to avoid frequent read and write disk operations. Data are written to the disk only when the cache is full or unchanged for a long time.
7. Terminals can be installed in parallel to guarantee reliable communication.

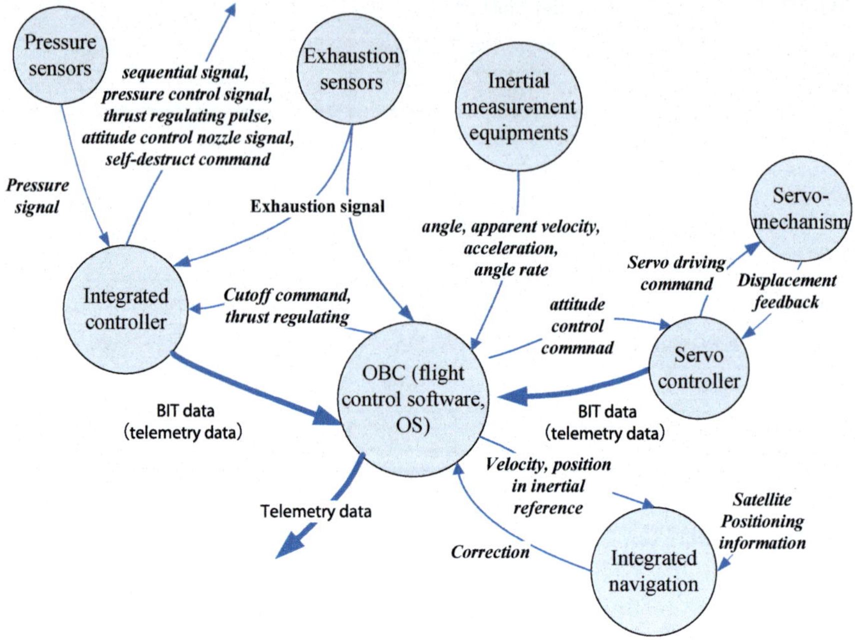

Fig. 7.12 OBS typical information flow

Figure 7.13 shows the BI block diagram, including a user mailbox for active communication convenience between the OBS and GS.

The functions of each block are briefly described as follows.

1. Bus inspector.
   The bus inspector monitors the processor state automatically, combines data from memory or I/O interfaces configured in advance into data units and transmits them to TX_FIFO.
2. Timer.
   The timer controls the transmission cycle, which can be set between 0 and 100 ms as required.
3. Transmission control.
   Transmission control includes two TX_FIFO queues in a ping pong mode based on the timer cycle. The FIFOs record or distribute data in turn to ensure data integrity and continuity. The process is automatic without FSW intervention.
4. Receiving control.
   Receiving control triggers communication interrupts and sets the corresponding state to inform the main processor to receive uploading data when they are delivered to the mailbox.
5. Mailbox control.
   Mailbox control actively transmits and receives data. For data not monitored by the inspector, flight software could deliver them to the mailbox and initiate a

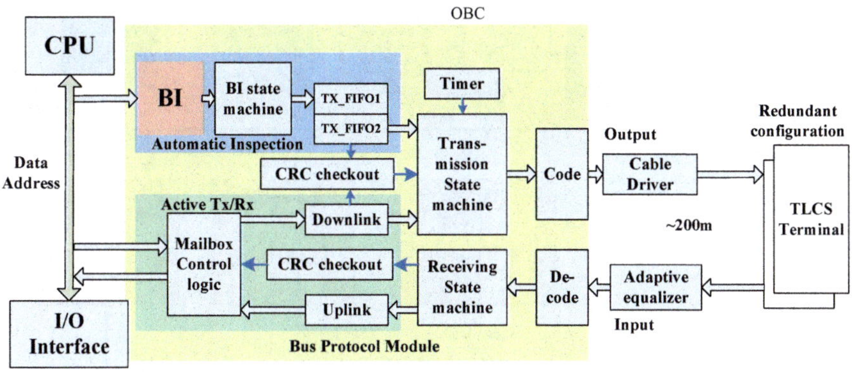

Fig. 7.13 Bus interception block diagram

download request through the transmission unit. This is a supplement to BI function, and GS commands can also be sent to the OBC through the uplink mailbox.

6. Code and decode circuits.

   Code and decode circuits pack transmitted data to protocol frames whenever FIFO and mailbox buffered data are not empty, and unpacks received data. Frame lengths are flexible, and the protocol accommodates varying length packets.

LVDS technology is appropriate for long distance and high-speed transmission, with reliability enhanced by 8b/10b encoding. Derating transmission is adopted with no less than 10 Mbps at 200 m. The BI automatically checks information based on the configured data addresses of the main processor, obtaining specific data from memory. Thus, FSW parameters can also be monitored. GS terminals can collect all necessary original OBS data instantaneously, which provides the basis for model-based automatic data analysis.

FSW design should accommodate this new test solution, adding a data management (DM) function as well as GNC processing, as shown in Fig. 7.14. A total of nine tasks are introduced, with priorities from high to low, as follows.

1. Initializing (INIT),
2. Timer interruption with 1 ms cycle (ISR_1).
3. Timer interruption with 20 ms cycle (ISR_20).
4. Communication interruption (ISR_comm).
5. Cutoff control (CC).
6. Main control (MC).
7. Mailbox (MB).
8. Data management (DM).
9. Other remaining OS tasks (TSK_os),

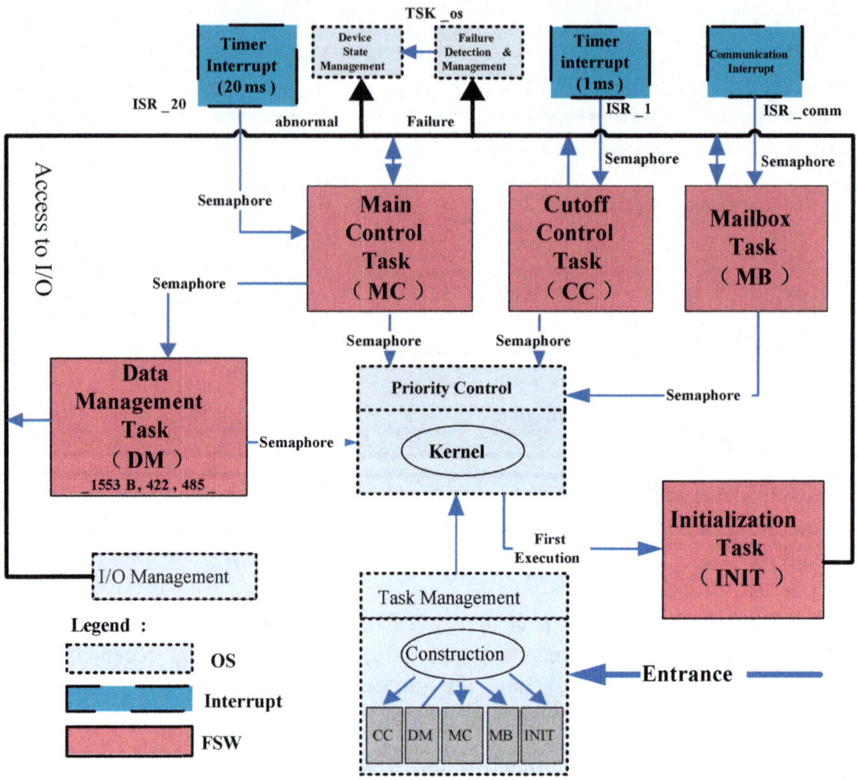

Fig. 7.14 Expanded flight software with embedded RTOS

DM has relatively low priority, so will not affect real-time control, collects BIT data from other remote terminals and relays them to telemetry system. These data are intercepted by the GS. All tasks are managed by the embedded RTOS.

The RTOS is not just applied for the test solution, but also benefits data management. Data volume is not the same in every control cycle, and RTOSs are more effective at dealing with peak data flow. For example, if the DM's task is to deal with a large data volume while the MC interruption task is triggered, since MC priority is higher than DM, the MC task will run immediately. Once it completes, the DM task continues. Thus, we can take full advantage of OBC runtime while ensuring higher priority tasks remain unaffected.

ICU software is also slightly modified, particularly for synchronization. OBS operation and GS analysis cannot be strictly synchronized, so all sensor data collected onboard are time tagged with reference to a start point, or time base, usually time to takeoff, engine shutdown, stage separation, etc. In some situations, e.g., filter algorithms used onboard, a data grouping indicator is also packed into the download message to assist analysis; otherwise, the data may be grouped differently on ground processing, causing calculated result inconsistent with that of the OBS.

7.3.3 *Data Analysis*

Depending on the LV control system characteristics, the data to be analyzed can be divided into three categories, as shown in Fig. 7.15.

1. Inertial measurement signals, e.g., IMU, rate gyro, and accelerator outputs.
2. Navigation and control parameters.
3. Sequential control signals, including pressure control signals, which can be characterized as discrete signals.

7.3.3.1 Inertial Measurement Signal Analysis

Inertial signals should be checked first, since they are the foundation for subsequent operations. For example, consider the system level testing environment. The IMU has been installed in the LV, so it is impossible to examine it following the methods used for calibration. Therefore, we propose three methods to verify IMU performance.

1. Comprehensive evaluation.
   Inertial measurement results are compared with Earth's rotational rate and local gravitational acceleration. This is a comprehensive assessment of multiple inertial meters, rather than attempting to distinguish outputs of each measuring axis (see Sect. 4.1.1.2).
2. Decomposition evaluation.
   Earth's rotational rate and local gravitational acceleration are decomposed into the theoretical outputs along each measuring axis of the IMU and compared with the measured value from the meters. Deviation is determined from the calibration result, leveling error, and IMU pointing error of the IMU (see Sect. 4.1.1.2).
3. Consistency evaluation.
   IMUs are usually redundantly configured for reliability. The output of one IMU can be compared to the other redundant IMUs. Since the IMUs are installed on the same holder, level and pointing error will be the same, and any error caused by mounting deformation may be ignored.

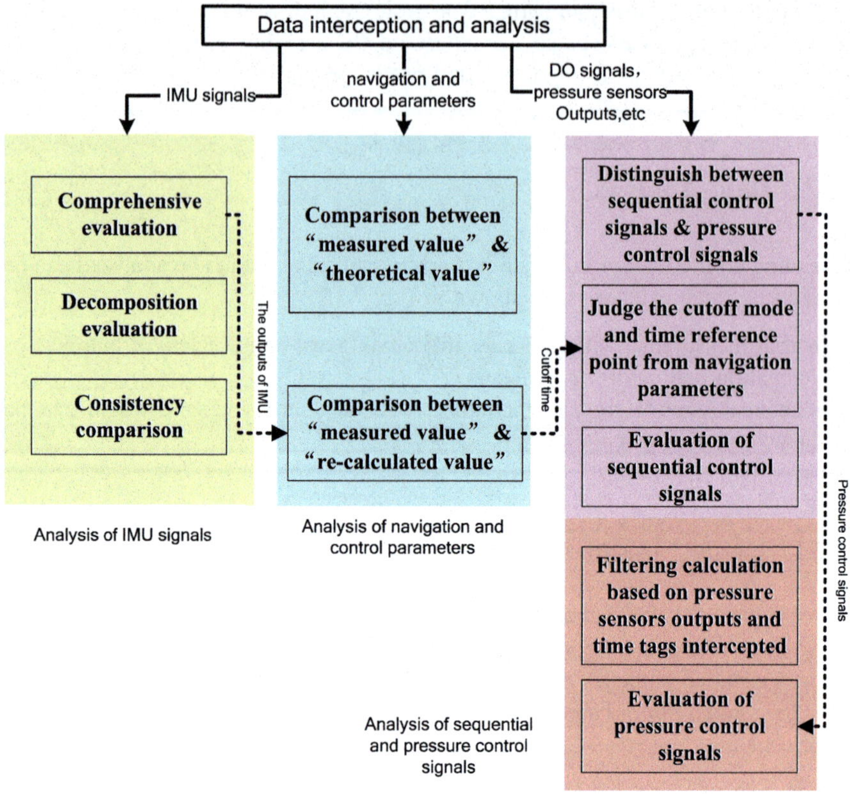

Fig. 7.15 Test data analysis process

For example, in TMR redundant design, deviations of attitude angle rate increments are

$$\begin{cases} \left| \sum^{T_g} \delta\theta_{x1} - \sum^{T_g} \delta\theta_{x2} \right| < \varepsilon_x \\ \left| \sum^{T_g} \delta\theta_{x2} - \sum^{T_g} \delta\theta_{x3} \right| < \varepsilon_x \\ \left| \sum^{T_g} \delta\theta_{x3} - \sum^{T_g} \delta\theta_{x1} \right| < \varepsilon_x \end{cases} \tag{7.1}$$

where T_g is the cumulative cycle, ε_x is the error threshold, $\delta\theta_{xi}$ (i = 1–3) is the angle rate output along the x axis, and i is the IMU identifier. There is a fault condition if any of the inequalities do not hold. Angle increments, apparent velocity increments, and apparent velocity can be examined using the same methods as FDI during flight, with more stringent error thresholds than the in-flight situation, since there is no vibration during the static test.

These methods are more rigorous compared to traditional functional evaluations, and the consistency judgment is unaffected by local gravitational acceleration, launch direction, or installation levelness. Thus, this constitutes an acceptable performance check for system level testing.

7.3.3.2 Navigation and Control Parameter Analysis

Two analysis methods are performed to check the navigation and control parameters calculated by the OBS based on credible inertial measurement signals.

1. Compare the measured and theoretical values.
   The measured value refers to the intercepted OBS results through the HMB. We can calculate the theoretical navigation and control parameter values on the ground terminal from known test site physical and geographical parameters and pointing azimuth using Runge–Kutta integration and nominal control equations.
2. Compare the measured and recalculated values.
   We can recalculate the navigation and control parameter values on the ground terminal using the intercepted IMU data (i.e., OBC inputs) and the same algorithm as in the onboard software, then compare them with the OBS values, i.e., the measured values.

7.3.3.3 Discrete Signal Analysis

There are two control command types: attitude control commands, which can be analyzed following the process above for control parameters (Sect. 7.3.3.2); and discrete commands, including SC commands determined by cutoff equations, and PC commands determined by real-time closed-loop pressure control algorithms.

The time base is critical for SC signals analysis. This provides the start point for each sequence, and all signals within the sequence have determined time intervals referenced to the time base. The time base is generally chosen as stage shutdown points, whose values vary in different cutoff modes, such as exhaust cutoff, guidance shutdown, timer termination, etc.

Pressure control is performed by ICUs installed in different rocket stages, and it is a little more complicated to analyze PC signals by reproducing the processes on the data handling terminal. Four threads are constructed in the terminal, as shown in Fig. 7.16. The left column describes how to determine PC strategies: time segments for PC permission or prohibition are updated based on the current time base, and real-time PC thresholds are calculated. The thresholds include upper limits (pressurization should be stopped if the pressure exceeds the limit), and lower limits (pressurization should be initiated if the pressure goes below the limit). Both

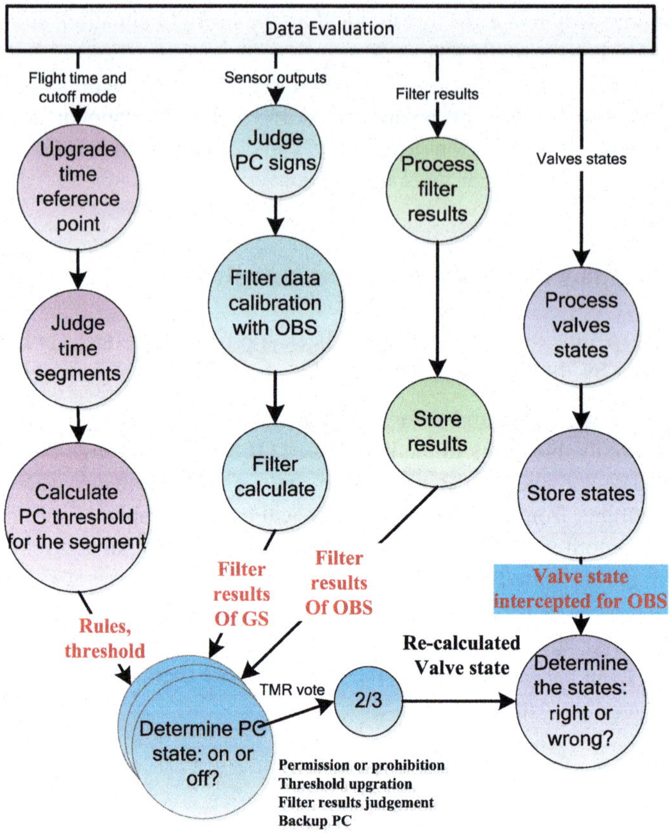

Fig. 7.16 Pressure control signal analysis

thresholds alter during flight and are calculated depending on real-time flight state and time base.

The second column describes the treatment for original pressure data collected from sensors and transmitted by BI. The data are filtered in group identical to the ICU process. For example, 10 samples are filtered as a group.

The third column describes ICU filter results, and is compared with GS calculated outputs as described in the second column. An error indication is sent if the comparison is inconsistent. Otherwise, PC state is determined based on the current time segment, thresholds, and filter outputs. If the filter outcome, i.e., pressure, exceeds some defined reasonable range, it is marked as untruthful and discarded, and a backup PC strategy is evoked. PC state is judged individually corresponding to every unit in a triple redundant ICU along with paired pressure sensors, where processes are the same for all the units and sensors. The final PC state is obtained using the two-out-of-three voting principle.

The right column describes the valve states. Valve state (on or off) sampled from the OBS is checked with the voted PC state described in the third column, as well as checking if a discrete PC command has been missed, accidentally triggered, or inconsistent in activated time, etc.

7.3.3.4 Failure Detection and Isolation

The data analysis system introduced above is more accurate and timely for failure isolation compared to traditional expert systems or IVHM. Figure 7-17 shows a simplified model of the control system.

IMU output is sampled by the OBC, where the flight software is running and transmitting attitude control commands to the SCU. After correction, filtering and decomposition, the SCU drives each servo mechanism to swing the engines. If there is an IMU failure, feedback of swing angle measured by the shift sensor would not meet the desired value and an error indication would pop up. Traditionally, experts would propose possible failure segments based on threshold judgment, shown as ①–⑩ in Fig. 7.17a, where segments ②, ⑥, and ⑩ represent communication cables between equipment. Failure in any segment could cause an unexpected swing feedback. If OBC and SCU outputs are also monitored, they will not match the predesigned values. Thus, appropriate error reports or indicators are produced. However, these reports or indications just warn there are system violations, and further work is required for failure isolation. It is difficult to identify the failure point due to the lack of necessary data from the limited test points.

The proposed failure isolation process is shown in Fig. 7.17b. With HMB and BIT, all data in each segment are intercepted and delivered to the GS. An IMU error would be identified first by the method of Sect. 7.3.3.1. Since IMU output is the same as the OBC sampled data (segments ② and ③), communication between the IMU and OBC and the OBC input interface, can be verified as operating properly. If the ground terminal software uses the intercepted original IMU outputs and

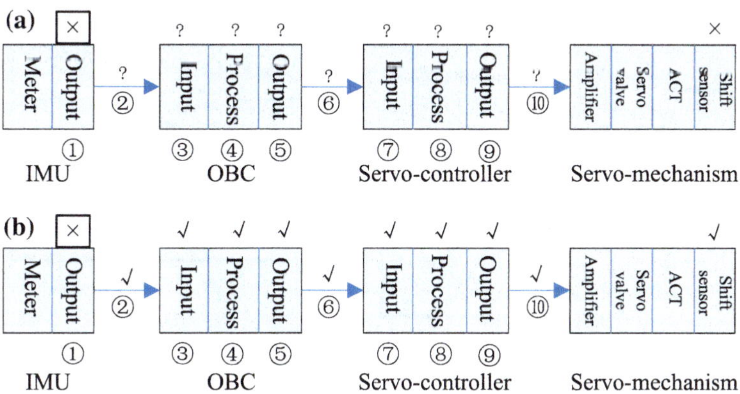

Fig. 7.17 Fault diagnosis process for **a** current and **b** proposed systems

obtains the same results as the OBC, then segment ④, flight software processing, can be verified. Subsequent segments can be verified until only segment ① remains and can be identified as the fault source.

In the case introduced above, although there are disparities between device actual outputs and expected values, these outputs correspond to the inputs and are verified by the playback on the GS. Thus, the search scope for failure points is greatly reduced.

An expert would also analyze the relationship between input and output of a device for FDI. Hence, the proposed process is superior to rule-based expert systems. Without BI, neither real experts nor expert systems can make an effective estimate with limited information.

7.4 System-in-the-Loop Test

With a high-speed downlink and uplink channel, a closed-loop test scenario can be performed, as shown in Fig. 7.10. We call this the system-in-the-loop test (SySIL) to distinguish from traditional semi-physical simulation tests. All equipment (rather than a subset) participate, and LV dynamics are simulated in the ground terminal and integrated into the launch control system. Figure 7.18 shows the SySIL testing workflow.

Simulation tests are usually only performed in the design department. However, SysIL would allow simulation tests to be integrated with system checks at the launch site or assembly building. This provides the advantages of system level-integration and simulation tests. All onboard hardware and software, including connections, maintain in-flight states, which is most important and acceptable to launch operators.

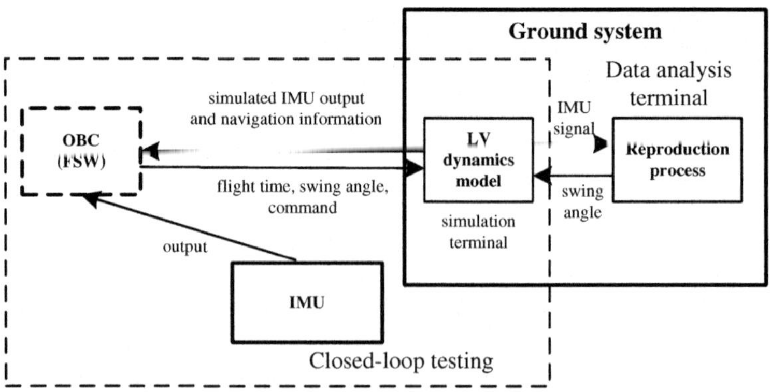

Fig. 7.18 Closed-loop testing workflow

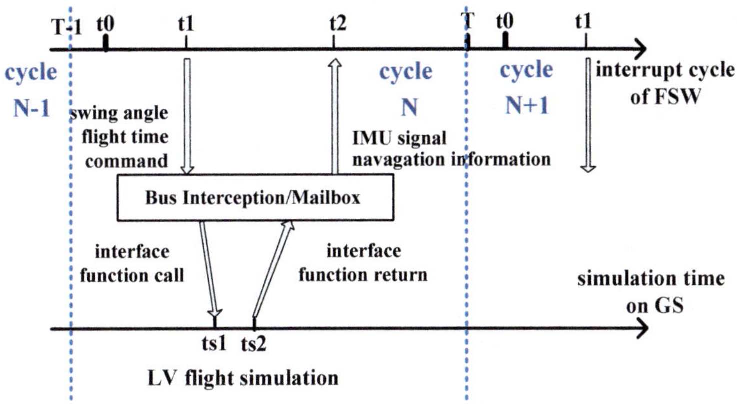

Fig. 7.19 Running time sequences for closed-loop testing

In this BI-embedded architecture, the LV dynamics model is running in the ground simulation terminal. The terminal obtains all the data for simulation, such as inertial navigation data, and FSW-calculated results. Then, it simulates the flight process as well as desired IMU outputs, and this simulated inertial signal will be fed back to the OBC through the uploading mailbox, forming a closed loop, which is essential to evaluate closed-loop guidance algorithms. This also enhances test coverage at the launch site, where test cases impose various disturbances and could be performed to examine flight control algorithm robustness until the final checkpoint before the flight. The running time sequences are shown in Fig. 7.19.

Starting from time t_0, the OBC samples servo swing angle, flight time, and control command at time t_1, and the simulation terminal receives the data simultaneously through BI. The simulated LV attitude, velocity, and acceleration are converted into IMU outputs and sent to the mailbox. At time t_2, the OBC obtains the mailbox data, continues this cycle of calculation and transmits the control command. The OBC operation continues through the next cycle until the test completes.

If the simulated data only replaced the real inertial signal in the test, the IMU could not be checked. Therefore, simulated and real IMU outputs are both sampled by the FSW as follows, where the simulated inertial signal is w_c, and the IMU theoretical output during testing is w_s.

1. The signal sent through mailbox by the ground simulation terminal is

$$w_m = w_c - w_s. \tag{7.2}$$

2. The real IMU output (w_i) can be regarded as the combination of the theoretical value (w_s) and random error, δ.

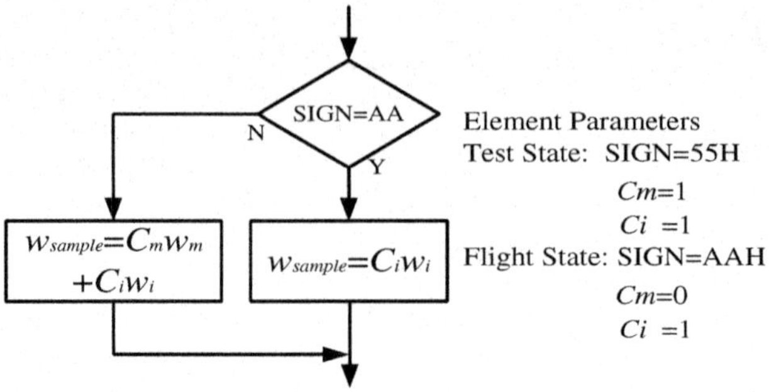

Fig. 7.20 Flight software reliability design

3. FSW samples the real and simulated signals,

$$w_{\text{sample}} = w_m + w_i = w_c - w_s + w_s + \delta = w_c + \delta \tag{7.3}$$

IMU random error is introduced in Eq. 7.3, but it is important to note that only static error can be checked, not dynamic error.

However, mailbox data should be excluded during flight. Although we can design different software versions for testing and flight scenarios, it violates the principle of *test what you fly*. Therefore, to ensure the same software for testing and flight, we use the reliability design shown in Fig. 7.20, with two weights, c_m and c_i, to control which inputs participate in the final sampling,

$$w_{\text{sample}} = C_m w_m + C_i w_i. \tag{7.4}$$

1. Different branches are controlled by a sign parameter. For example, 55H means testing state, and the OBC samples the mailbox and IMU data; AAH means flight state, and the OBC only samples IMU data.
2. If the software runs the wrong branch while in flight (the "N" branch), since $C_m = 0$ in the flight state software, there is no influence on the outcome.
3. During flight, mailbox communication is prohibited and all data in the mailbox memory = 0.

Analysis criteria need not be prepared in advance for each SySIL test case, but can be simultaneously generated in the GS data handling terminal. Analysis granularity is greatly increased since data are inspected every control cycle. These aspects are welcomed by test staff, and on-site technical support personnel are greatly reduced. TLCS front-end equipment configuration can also be greatly simplified, as shown in Fig. 7.21 compared with Fig. 6.1 in Sect. 6.1.2:

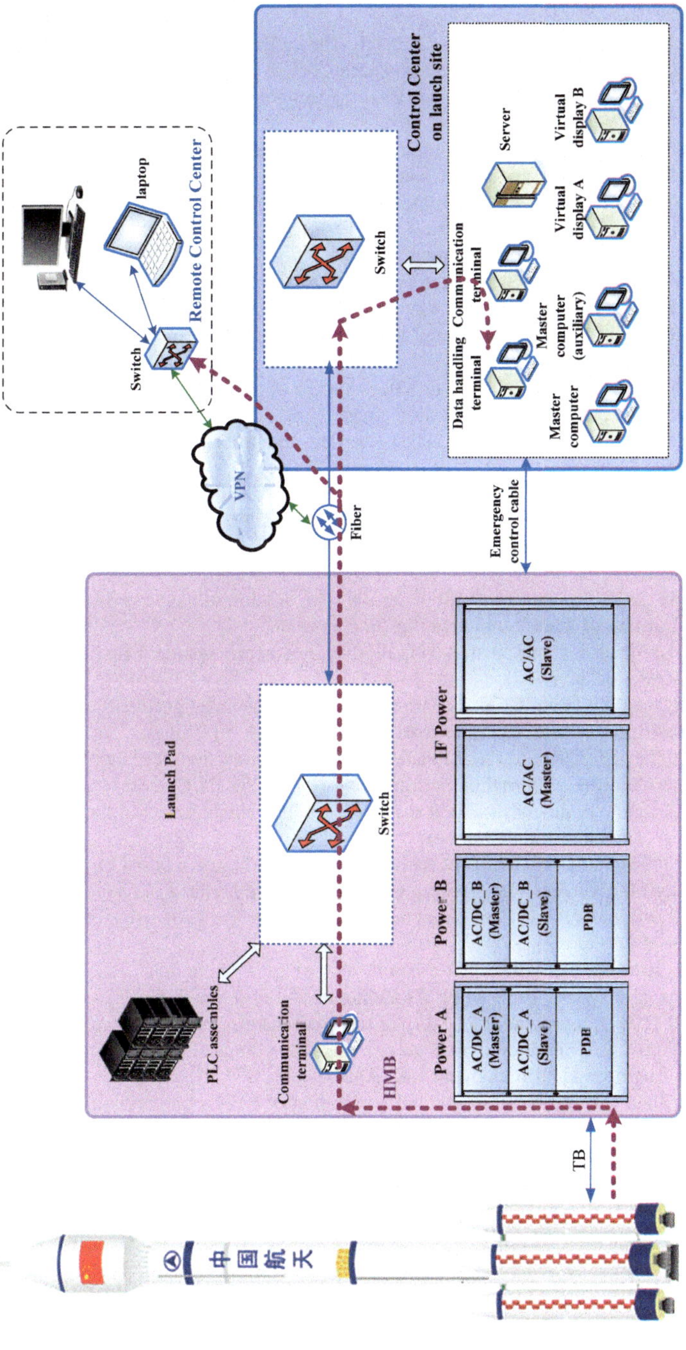

Fig. 7.21 Simplified launch pad configuration with the proposed system (compare with Fig. 6.1, Sect. 6.1.2)

References

1. Song ZY (2013) Smart launch, free fly-The effect of control system on elevating competitiveness of launch vehicle. Aerospace Control 31(6):3–9,43 (in Chinese)
2. Zhou JP (2013) Space rendezvous and docking technology. National Defense Industry Press, Beijing (in Chinese)
3. Song ZY (2013) The development trend of test and launch control technology on future space transportation system. Aerospace Control 31(4):2–10 (in Chinese)
4. Huang JB, Zi XY, Liu GP (2006) Design and implementation on detonator testing unit of some missile. J Proj, Rocket, Missiles And Guid 26(2):393–395 (in Chinese)
5. Zhou H, Song ZY (2013) The design and research of closed loop general testing technique based on BMU and LVDS bus. Aerospace control 31(3):45–49 (in Chinese)
6. Patterson-Hine A, Narasimhan S, Aaseng G, et al (2005) A review of diagnostic techniques for ISHM applications. 1st Integrated Systems Health Engineering and Management Forum, Napa, USA, November 7–10
7. Jiang LX, Li HW, Yang GQ et al (2009) A survey of spacecraft autonomous fault diagnosis research. J Astronaut 30(4):28–34 (in Chinese)
8. Long B, Song LH, Jing WX et al (2003) A review and propect of fault diagnosis technique of spacecrafts. Missilles Space Veh 3:31–36 (in Chinese)
9. Schwabacher M, Waterman R (2008) Pre-launch diagnostics for launch vehicles. In: Proceedings of the IEEE Aerospace Conference, Big Sky, USA, March 1–8, 2008
10. Cai ZX (2004) Intelligent control, 2nd edn. Publishing House of Electronics Industry, Beijing (in Chinese)
11. Wang ZH (2010) Research on fault diagnosis approach for attitude control system of satellite. Harbin Institute of Technology, Harbin (in Chinese)
12. He GQ, Liu JY et al (1998) A missile faullt diagnosis expert system. J Beijing Inst Technol 18 (6):685–688 (in Chinese)
13. Chen LL, Liu CR, Ren Z (2008) Knowledge base system designed for launch vehicle fault diagnosis expert system. Missilles Space Veh (6):11–15 (in Chinese)
14. Liu CR, Zhang QZ, Ren Z (2008) Fault diagnosis expert system for launch vehicle based on augmented fault tree. Journal of Astronaut 29(6):258–263 (in Chinese)
15. Xia Y (2007) Fault diagnosis expert system of launch vehicle based on fault tree. Chongqing University, Chongqing (in Chinese)
16. Jin LL (2008) Study on spacecraft fault diagnosis expert system based on fault tree. Nanjing University of Aeronautics and Astronautics, Nanjing (in Chinese)
17. Watson I, Marin F (1994) Case-based reasoning: review. The knowledge Engineering Review 9(4):327–354
18. Aamodt A, Plaza E (1994) Case-based reasoning: foundational issues, methodological variations, and system approaches. AI Communications 7(1):39–59
19. Heider R (1997) Improving the quality of technical data for developing case based reasoning diagnostic software for aircraft maintenance. In: Proceedings of 13th International Conference on Data Engineering, Birmingham, UK, April 7–11
20. Yang J, Zhao QY (2008) Research and application of CBR's progression. Comput Eng Des 29(3):710–712 (in Chinese)
21. Yoshimoto B (1995) Vertical launching system (VLS) technical assistance expert system (VTAEXS). In: 11th Conference on Artificial Intelligence for Applications, Los Angeles, USA, February 20–23
22. Pepe C, Beltz S, Merchant R, et al (1994) KATE: From the lab to the firing room. NASA, NASA-CR-201411
23. Parrish CL, Brown B (1991) Knowledge-based autonomous test engineer (KATE), NASA, N92-22685
24. Hou X (2000) The control system's fault diagnosis system of the launch vehicle. Aerospace Control 18(1):67–72 (in Chinese)

25. Fu WX, Zhu SP, Yan J et al (2007) Fault diagnosis for power system of carrier rocket using parameters estimate algorithms. J Proj, Rocket, Missiles And Guid 27(1):181–183 (in Chinese)
26. Kobayashi T (2003) Aircraft engine sensor/actuator/component fault diagnosis using a bank of Kalman filters. NASA, NASA/CR-2003-212298
27. Falcoz A, Henry D, Zolghadri A (2010) Robust fault diagnosis for atmospheric reentry vehicles: A case study. IEEE Trans Syst Man Cybern A Syst Hum 40(5):885–899
28. Patton RJ, Chen J (1997) Observer-based fault detection and isolation: robustness and applications. Control Engineering Practice 5(5):671–682
29. Wells SR (2002) Application of sliding mode methods to the design of reconfigurable flight control systems. University of California Davis, California
30. Zhang J (2009) Evaluation of observer structures with application to fault detection. Northeastern University, Massachusetts
31. Chen XQ, Geng YH, Zhang SJ, et al (2007) A mixed H2/H∞ approach to integrated fault diagnostic and fault-tolerant control. J Astronaut 28(4):124–130 (in Chinese)
32. Gorinevsky D, Samart S, Bain J, et al (2005) Integrated diagnostics of rocket flight control. In: IEEE Aerospace Conference, Big Sky, USA, March 5–12
33. Frank PM (1990) Fault diagnosis in dynamic-system using analytical and knowledge-based redundancy: a survey and some new results. Automatica 26(3):459–474
34. Zhou DH, Liu Y, He X (2013) Review on fault diagnosis techniques for closed-loop systems. Acta Automatica Sinica 39(11):1933–1943 (in Chinese)
35. Kurtoglu T, Johnson SB, Barszcz E, et al (2008) Integrating system health management into the early design of aerospace systems using functional fault analysis. In: International Conference on Prognostics and Health Management, Denver, USA, October 6–9
36. Chandola V, Banerjee A, Kumar V (2009) Anomaly detection: a survey. ACM Computer Surveys, 41(3):15:1–15:58
37. Liu XM, Li H, Jiang JB (2010) A diagnostic method for launch vehicle based on fault tree and neural Networks. Computer Simulation 27(7):38–42,77 (in Chinese)
38. Bay S D, Schwabacher M (2003) Mining distance-based outliers in near linear time with randomization and a simple pruning rule. In: Proceedings of the 9th ACM International Conference on Knowledge Discovery and Data Mining, Washington, USA, August 24–27
39. Ferrell B, Oostdyk R (2010) Modeling and performance considerations for automated fault isolation in complex systems. In: IEEE Aerospace Conference Proceedings, Big Sky, USA, March 6–13
40. Lv ZJ, Chen Z, Lv YH et al (2005) Research on method of fault diagnosis in control system based on fuzzy neural network. Aerospace Control 23(4):5–9 (in Chinese)
41. Yin MJ (2011) Research and implementation of launch vehicle fault diagnosis system. University of Electronic Science and Technology of China, Chengdu (in Chinese)
42. Zhu DQ, Liu YA (2007) Information fusion method for fault diagnosis. Control Decis 22 (12):1321–1328 (in Chinese)
43. Gros XE, Lowden DW (1995) Bayesian approach to NDT data fusion. Non destr Test Cond Monit 37(5):462–468
44. Bogler PL (1987) Shafer-dempster reasoning with application to multisensor target identification system. IEEE Trans Syst Man Cybern 17(3):968–977
45. Greensmith J, Aickelin U, Tedesco G (2010) Information fusion for anomaly detection with the dendritic cell algorithm. Inf Fusion 11(1):21–34
46. Lv F, Wang XQ, Du HL, et al (2009) Research state and progress of fault diagnosis based on information fusion. J Huazhong Univ Sci Tech (Natural Science Edition) 37(S1):217–221 (in Chinese)
47. Xing Y, Wu HX, Wang XL et al (2003) Survey of fault diagnosis and fault-tolerance control technology for spacecraft. J Astronaut 24(3):221–226 (in Chinese)
48. Gross AR, Gerald-Yamasaki M, Trent RP (2009) Analysis of space shuttle ground support system fault detection, isolation, and recovery process and resources. NASA, NASA/TM-2009-215406

49. Johnson SB, Gormley TJ, Kessler SS (2011) System health management: with aerospace applications. John Wiley & Sons Ltd, Hoboken
50. Fox JJ, Glass BJ (2000) Impact of integrated vehicle health management (IVHM) technologies on ground operations for reusable launch vehicles (RLVs) and spacecraft. In: IEEE Aerospace Conference Proceedings, Big Sky, USA, March 18–25
51. Johnson SB (2005) Introduction to system health engineering and management in aerospace. In: 1st Integrated Systems Health Engineering and Management Forum, Napa, USA. November 7–10
52. Saxena A, Roychoudhury I, Lin W, et al (2013) Towards requirements in systems engineering for aerospace IVHM design. In: AIAA@Infotech 2013, Boston, USA, August 19–22
53. Datta K, Jize N, Maclise D, et al (2004) An IVHM systems analysis & optimization process. In: IEEE Aerospace Conference Proceedings, Big Sky, USA, March 6–13
54. Edwards J, Farner S, Gershzohn G R, et al (2007) Space transportation—Delta launch vehicle system health monitoring, failure detection, diagnosis and response. AIAA Space Conference & Exposition, Long Beach, USA, September 18–20
55. Poll S, Iverson D, Ou J, et al (2003) System modeling and diagnostics for liquefying-fuel hybrid rockets. NASA, NASA/TM-2003-212270
56. Iverson DL (2004) Inductive system health monitoring. In: Proceedings of the International Conference on Artificial Intelligence, IC-AI '04, Las Vegas, USA, June 21–24
57. Biswas G, Mack D, Mylaraswamy D, et al (2013) Data mining for anomaly detection. NASA, NASA/CR-2013-217973
58. Zhang SM, An XY, Yan TG et al (2013) Analysis and discussion of health management technology for large launch vehicle. Missiles Space Veh 6:33–38 (in Chinese)
59. Bo W, Cai YW, Tong J et al (2009) Analysis of impact from IVHM to China's carrier rocket and testing and launching control system. Ordnance Industry Automation 28(1):78–80 (in Chinese)
60. Song ZY (2015) Data-driven responsive test technique for new generation launch ehicle. J Astronaut 36(12):1435–1443 (in Chinese)

Printed by Printforce, the Netherlands